Preparing to Measure Outcomes

Preparing to Measure Outcomes

**A GUIDE FOR DEVELOPING
QUALITY ASSURANCE IN THE
HUMAN SERVICES**

*Phyllis M. Johnson,
Teresa L. Kilbane,
and Laura E. Pasquale*

CWLA PRESS
WASHINGTON, DC

CWLA Press is an imprint of the Child Welfare League of America. The Child Welfare League of America is the nation's oldest and largest membership-based child welfare organization. We are committed to engaging people everywhere in promoting the well-being of children, youth, and their families, and protecting every child from harm.

CHILD WELFARE LEAGUE OF AMERICA, INC.
HEADQUARTERS
440 First Street, NW, Third Floor, Washington, DC 20001-2085
E-mail: books@cwla.org

CURRENT PRINTING (last digit)
10 9 8 7 6 5 4 3 2 1

Cover and text design by Jennifer R. Geanakos
Edited by Julie Gwin

Printed in the United States of America

ISBN # 0-87868-850-1

Library of Congress Cataloging-in-Publication Data
Johnson, Phyllis M., 1941-
    Preparing to measure outcomes : a guide for developing quality assurance in the human services / Phyllis M. Johnson, Teresa L. Kilbane,
Laura E. Pasquale.
        p. cm.
Includes bibliographical references and index.
    ISBN 0-87868-850-1 (alk. paper)
    1. Human services--Evaluation. 2. Evaluation research (Social action programs)
3. Quality assurance. I. Kilbane, Teresa L. II. Pasquale,Laura E. III. Title.
HV40.J65 2003
361'.0068'4--dc22   2003023224

CONTENTS

LIST OF TABLES

Foreword

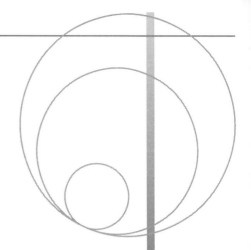

Several years ago, I realized that agencies needed a guidebook on quality. At the time, I was designing a quality program for a large private, nonprofit child welfare organization. Although everyone in the organization wanted the new quality program, communicating the concepts was very difficult. Everyone involved had different understandings of terms that relate to "quality" and what the terms required us to do.

The 1980s, the decade of reengineering and rightsizing, ushered in articles and books on quality management. U.S. corporations were losing revenues to businesses in other countries because of poor product quality. Business losses brought on a new trend of rapid and complex change. As the 1980s drew to a close, constant change typified the work environment. Organizations and employees in both the for-profit and nonprofit sectors were forced to contend with rapid change.

In the 1990s, management gurus such as Darryl Connor recognized the challenge of change and encouraged us to manage speedily. Other learning organization masters, such as Peter Senge and Chris Argyris, urged us to sharpen our perceptive skills so we could think our way through new situations, adapt, and work as quickly as possible. These experts told us that we must be in "learning and doing modes" simultaneously, because we could no longer afford to take our time.

Change is inevitable. At the same time, the forces that precipitate change often come from outside the organization and, therefore, are out of the organization's control. Change can have negative effects on organizations and their employees. Employees must have tools to help them understand change, incorporate that understanding into work routines, and perform well as quickly as possible.

The challenges we face now, and the ways most human service organizations have continued to operate, are representative of the past. In the past, change was infrequent, slow, incremental, and usually confined to small areas. Employees were not to deviate from prescribed routines. Because workers relied heavily on long-held work patterns, they were not concerned about what occurred in other departments. Most employees gave little thought to what happened before or after their tasks in the work process. When problems developed in projects requiring combined efforts, the usual reply was, "I did my part." Only a few employees, those at the bottom and at the top, had to be concerned about customer satisfaction. As a consequence, little attention was paid to the quality and effectiveness of the employees and processes between the two ends of the organizational hierarchy.

Today, waves of change are frequent and swift, usually affecting the entire organization. When change occurs, old routines do not work. Workers must have goals, objectives, and outcomes in mind, and they must possess a wide range of skills so they can stay on course to the expected outcomes. A high level of collaboration is required. Value is placed on knowledgeable and informed work and the capitalizing of front-line wisdom. All levels and types of employees must worry about customer satisfaction. Support services within the organization must even be concerned about the satisfaction of internal customers.

Difficulties in detecting organizational problems produce the greatest stumbling block. In the past, experts in organizational development relied on time-proven theories for developing solutions to organizational problems. In contemporary organizations, however, old solutions do not work, and the shelf life of any solution is short.

In my work, I have tried to interpret ideas from organizations with which I have worked. I have encountered stark differences between my interests and those of other staff. In most cases, the organizations' staff had few learning experiences that helped them connect the concepts of a full-quality system with their human service professional practice. In fact, many of my colleagues considered quality-related activities "extra work" and not part of their jobs. As I faced the task of going beyond the problem-finding activities of traditional quality assurance, I began with established objective processes typical of the health care quality field. I found, however, that three problems arose when objective evaluations or reviews were used alone.

First, identifying the problems (the purpose of objective reviews) does not eliminate them. Objective review helps individuals perceive mistakes, but the relationship between the mistakes and the individuals' actions are not always understandable to them. The employees are often left to figure out how to correct the mistakes. Second, objective evaluations allow workers to continue thinking about quality as something that others do, something that is not part of their responsibility. This mindset is one of the most difficult cultural barriers to overcome. Third, if only objective processes are used in the organization's quality system, administrators and managers can easily distance themselves from accepting responsibility for the organization's outcomes. Managers may see their own failures solely in terms of not having applied sufficient pressure to ensure that employees do whatever is needed to produce the desired results.

To be successful, organizations must increasingly rely on the preparation and capabilities of teams of front-line workers, not managers. Two skills— thinking strategically and communicating effectively—have moved from options to fundamental requirements. In the past, thinking skills were considered inborn traits that were only marginally modifiable. The assumption was that highly intelligent people did not have to improve their thinking and that individuals who had to struggle intellectually could not improve. It is now recognized that most individuals can improve their thinking skills.

Similarly, communicating effectively is an essential skill in modern organizations. Many organizations cite poor communication as a major reason for organizational problems. For example, in agencies in which administrators have changed the titles of departments and positions, it is not uncommon to find that staff do not agree on the titles of the departments or the appropriate position labels. Unfortunately, staff also disagree on the tasks they are required to perform and have very different perceptions of expectations and desired outcomes.

Another contrast between the past and present that often is not addressed is work pace, that is, the tempo at which job tasks are performed. Time parameters in the current environment are tighter than in the past. Although most of us are aware that we have more work these days, we have not made the necessary adjustments on the job. The pace of jobs is defined by the organization's culture. Most employees quickly determine the accepted pace and work at a pace that is not too different from fellow employees. Although organizational leaders may ask for more output and better outcomes, the organizational culture reinforces the slower work paces that were effective in the past. Even in the present environment, many employees continue to pace the completion of tasks within the confines of out-of-date rhythms. Today, because work processes are so interconnected, the pace of work across processes must be fairly well in sync. Employees must be helped to perceive quicker paces as culturally acceptable, with peers reinforcing the work rates necessary for the new, faster environment.

As I incorporated self-awareness and self-assessment tools in my work with social work staff, I found that employees began to become enthusiastic about participating. Clearly, many lacked the skills required to perform quality activities, as well as the knowledge of how and why to apply those skills. Many have been unclear about what the skills would be used for, which tasks comprised quality work, and the order in which to perform tasks for completing quality processes. In informal conversations, in focus and discussion groups, and by surveys and interviews, I have asked many child welfare professionals what they needed to help them incorporate quality techniques into their work. "Some kind of book or manual" is the most frequent answer. I now recognize that not only are human services still in "then," modes—that is, they are relatively unfamiliar with the concepts and tenets of contempcrary organizations—but professionals-in-training generally are not being educated on the importance of incorporating quality-related practices into their work. Although research methods are still required courses in most schools of social work, they provide skills only in identifying problems, which is only half of the quality equation. (Besides, most social workers consider research courses punishment.) To be helpful, a quality system must guide employees to change error-producing habits into quality-promoting practices that achieve desired outcomes.

New researchers are framing the new imperatives for human service organizations and the people who work in them. I hope that *Preparing to Measure Outcomes* is a contribution to this work.

—Phyllis M. Johnson

Introduction

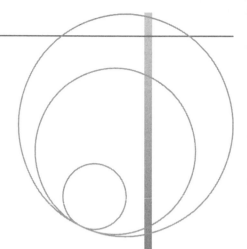

Preparing to Measure Outcomes leads users through the performance of reviews and studies to implement the evaluation component of an organization's quality system. Quality assurance is defined as an objective review of an organization's products and services to

- determine if work is being performed according to plan, and if expected outcomes are consistent with the plan; and

- provide data to the organization's leadership, so that the strategic plan can be changed to meet expectations.

Quality Functions

This guidebook presents a complete quality system that become part of an organization's overall strategic plan and integral to every unit's tactical plans. In this system, quality functioning is interwoven into the organization's values and culture, so that the organization is characterized by continuous improvement and excellence. At the core of such a quality system are three quality-based functions: management, assurance, and improvement. These functions have become important because the demands on modern organizations differ from those of less than two decades ago. Current demands require that individuals adjust smoothly to changing conditions, that organizations successfully grapple with access to fewer resources, and that employees do more work in a shorter time. These demands have made past standards of work acceptability inappropriate.

Quality management establishes new, more appropriate values; policies to create an organizational environment supportive of the new values and standards; and leadership that maintains the environment without impediments to organizational processes. Quality assurance teaches personnel how to incorporate these new standards into work tasks, makes clear when standards have been met, and helps workers develop ways of meeting the new standards. After personnel judge work based on the new values and standards, quality improvement extends evaluative tactics into peer supports, which helps staff identify their mistakes in work processes and correct them as early as possible.

Each of these three quality functions is important to the effectiveness of an organization's quality system. The prominence of each function in the organization and the timing for implementing each in a new quality system, however, are seldom the same for any two organizations. The first function, quality management, provides the foundation for the other two functions. As a result, it must be in place and well-established before quality assurance and quality improvement can be successful. Quality management includes ensuring the availability of tools and other resources for working, the requisite knowledge and skills on the part of staff so that they know what to achieve and how to go about it, and the prevention or removal of barriers that hinder achievement of desired outcomes.

Although quality assurance and quality improvement may be implemented in close succession, quality assurance must come second and, in the early stages, it will be more prominent than quality improvement. As personnel begin incorporating the new standards into their work routines, however, quality improvement will become the stronger and more prominent of these functions. Balance is achieved by emphasizing quality assurance and quality improvement equally. Too much emphasis on either function puts an unnecessary strain on the organization's resources; too little emphasis jeopardizes organizational quality.

The Preparation and Planning Phase of Quality Assurance

Quality systems are complex. Knowledge and skill are necessary to ensure their proper functioning. Quality assurance, the evaluative component of the system, involves several phases and multiple steps in each phase. Table 1 outlines the quality assurance phases, but only the first phase, Phase A **[WHY NOT PHASE 1?]**, is covered in this guidebook. Because Phase A involves preparation and planning, it is critical. If preparation and planning are inadequate, the data, results, and findings of the project could be worthless.

The quality system will function smoothly only if each function is defined by a plan. The quality plan details the types of strategies that will be used to verify that processes are functioning as intended and that outcomes are being achieved. The plan specifies how the organization will support staff to achieve expected outcomes (quality management), how personnel can support themselves and each other so expected outcomes are achieved (quality improvement), and how the organization will verify that processes are being performed and outcomes are being achieved as expected (quality assurance). In relation to each of the three functions, the quality plan

- defines the requirements to which the organization is committed,

- prioritizes the requirements,

- specifies the types of projects best suited to assess the processes and outcomes for each requirement,

- indicates how frequently each evaluation project should be completed, and

- identifies who should receive the reports of the results and the format of the reports.

Quality Assurance Methodologies

Quality assurance methods are based on research and evaluation to identify quality issues and problems. Once the company identifies issues and problems, administrative and management strategies are used to make improvements. The skills needed to do quality work come from both research and management (see Appendix A). Many organizations are more effective in identifying problems (research skills) than implementing solutions and making improvements (management skills). Management abilities, however, are as critical to quality success as research methods.

A number of evaluation designs can be used in quality assurance projects. The methods differ not in data collection methods, but in the specific purpose of the project, the rigor with which findings are interpreted, and the planned use of the results. Table 2 outlines evaluations in two categories—process and outcome—and describes the purpose of each type of evaluation.

One type of process evaluation is monitoring. Monitoring is a review to find out whether clients receive the services specified in their service plans. Monitoring may assess such issues as whether the service plan was completed with the family and whether contacts were made with or in behalf of the child or family. Customer satisfaction surveys or interviews with families to obtain their opinions of the services they received also monitoring studies, and they examine service delivery from the customer's perspective, validate case documentation, and verify other records of service receipt. Monitoring studies also include the collection and analysis of client appeals and grievances and studies designed to learn whether staff are performing tasks that lead to expected outcomes.

TABLE 1: Phases of the Quality Assurance and Evaluation Process	
A	Preparing to review, evaluate, and study
B	Implementing data collection: Conducting the review and evaluation
C	Making qualitative judgments and communicating them to reviewees, quality group members, and others
D	Guiding the development of individual improvement and action plans
E	Implementing individual improvement and action plans
F	Assessing effectiveness of individual improvement and action plan implementations
G	Analyzing and reporting to plan organizational improvement
H	Planning and implementing organization-wide improvement
I	Assessing progress of the organization-wide improvement plan

Two types of outcome evaluation are impact assessments and utility assessments. These evaluations assess whether expected outcomes are being achieved and how observed outcomes compare with expected outcomes. They also assess whether an intervention brought about any unexpected negative consequences or unplanned benefits. All outcomes research must be experimental or quasiexperimental, because the researcher makes conclusions about causation. Outcomes research, consequently, is a much more rigorous process than monitoring. Appendix B outlines the steps in conducting an outcome evaluation.

Quality Roles

Quality functions will be effective only when clear, correct information is communicated among personnel at all levels and in all positions of the organization. Quality functions require respectful workplace relationships and trust to ensure that information is mutually understood. Likewise, formal relationships must support outcome achievement. This view of organizational relationships is rooted in leadership and management philosophies and must be modeled routinely in everyday practice.

To establish the quality aspects of an organization, formal job tasks can be thought of as quality roles. There are nine essential quality roles: developer-manager, reviewer, data analyst, quality leader, quality group member, quality collaborator, quality advocate, reviewee, and coach. In some organizations, a single individual may effectively play more than one role, and in other organizations, more than one person may be needed to ensure that a role is covered adequately. Most important, the organization must ensure that each role is understood and that the tasks of all the roles are performed.

Quality professionals who perform quality assurance tasks must fill the roles of developer-manager, reviewer, and data analyst. Often, these roles are the professionals' primary job duties. Personnel with other primary duties may fill the remaining six roles. These roles—quality leader, quality group member, quality collaborator, quality advocate, reviewee. and coach—are not covered in this present text but are featured in the next volume and are central to an organization's quality managment and quality improvement processes. A complete quality program includes all three functions: quality assurance, quality management, and quality improvement. Every employee, however, is expected to assume the roles of reviewee and coach at least once as a member of the organization.

TABLE 2: Evaluation Types and Purposes

	TYPE	PURPOSE
Evaluation of Process	Comprehensive evaluation	Research and analysis on the conceptualization and monitoring of program interventions, and the assessment of program utility.
	Conceptualization and design analysis	Studies of the extent and location of target problems of intervention, ways to operationally define target populations, and the suitability of the proposed intervention.
	Fine-tuning evaluation	Evaluations to examine the implementation, effectiveness, and efficiency of modifications in existing programs.
	Monitoring	Assessing the extent to which a program is consistent with its implementation plan and directed at the appropriate population.
Evaluation of Outcomes	Impact assessment	Evaluation of whether and to what extent a program causes changes in the desired direction among a target population.
	Utility assessment	Study of the effectiveness or efficiency (cost-to-benefit ratio or cost-effectiveness) of programs.
	Cost-benefit analysis	Studies of the relationship between project costs and outcomes in monetary terms.
	Cost-effectiveness analysis	Studies of the relationship between project costs and outcomes in costs per unit of outcome achieved.
	Efficiency evaluation	Analyses of the costs (inputs) of programs in comparison to either their benefits or to their effectiveness (outputs). Benefits, like inputs, are expressed in monetary values; effectiveness is measured in terms of substantive changes in the behavior that is the program focus.
	Comprehensive evaluation	Research and analysis of the design of interventions, the monitoring of program interventions, and the assessment of program utility.

Source: Rossi and Freeman (1993).

Developer-Manager

The developer-manager designs and develops the quality system in conjunction with the organizational leadership and oversees implementation of quality plans, reviews and evaluations, and reporting of results. The developer-manager also recommends strategies for improvement and evaluates improvement projects for success.

Developer-managers must be highly skilled in all aspects of quality systems. They must use a number of strategies to collect the data required to perform the problem-finding tasks of the quality plan. Some of these strategies are

- reading case files to assess the worker's and unit's performance, including the completeness of case record documentation;

- reviewing appeals to identify any patterns that shed light on service delivery effectiveness;

- surveying service-related populations, such as foster parents, to assess whether service delivery is adequate;

- analyzing administrative and other data to assess the performance of the organization or of individual teams in achieving expected outcomes;

- visually surveying the physical facility to ensure that the environment facilitates job performance; and

- holding focus groups to capture information, define newly emerging quality issues, and develop plans for improvement before problems become ingrained in the culture.

Reviewer

In this book, the terms *reviewer* and *evaluator* are used interchangeably. Which term your agency uses depends on the purpose of the project. In addition, *data collector* encompasses both *reviewer* and *evaluator*. The reviewer reviews and evaluates organizational processes and outcomes. Activities include collecting and collecting and analyzing data and reporting the results of the review. Reviewers must intergrate information, reach sound conclusions, and synthesize their conclusions to make recommendations and help staff take actions that lead to the achievement of expected outcomes.

Data Analyst

The data analyst processes, analyzes, and reports data collected during reviews and evaluations in formats that help developer-managers and reviewers perform their duties effectively.

Quality Leader

The quality leader is a supervisor who makes the organization's values and vision clear by incorporating the organization's mission and goals in his or her communications. The quality leader is skilled at communicating support and has much of the knowledge and many of the skills identified as important for success in contemporary organizations. Leaders function at all organizational levels, not only at the top of the management hierarchy. High-level leaders, however, are the beacons that reinforce organizational values, help mid-level leaders set the right tone, and discourage behaviors that are barriers to achieving outcomes.

Quality Group Member

The quality group member belongs to a quality committee or team formed to respond to specific organizational issues. Such councils or teams may cover different geographic areas, be organized around different functions, or target different problems and issues.

Quality Collaborator

The quality collaborator works to clear the barriers and impediments that prevent the achievement of quality outcomes. Quality collaborators are skilled in quality management. They know the relationship between process alignments and achievement of outcomes and how to correct misalignment. They also motivate personnel to change their work routines. *Process alignment,* a term that has been in the management lexicon for several years, refers to how well coordinated are the work and organizational processes.

Quality Advocate

The quality advocate supports quality in practice by ensuring that the organization's strategic and tactical planning and policies enable (and do not hinder) necessary quality processes. Quality advocates must have a general knowledge of the scope of the quality system and possess the skills to enable the system to flow unfettered. Advocates also understand how quality systems are integral to effective practice.

Reviewee/Coach

The reviewee has his or her work reviewed, either as part of a random sample or to determine the individual quality of his or her practice and outcomes. Depending on an organization's structure and functions, reviewees may include direct service workers, supervisors, and middle- and high-level managers, such as executive directors, chief financial officers, chief information officers, and chief operating officers.

The coach helps colleagues integrate new behaviors into work processes by reminding coworkers when their actions conflict with the action plan. Coaches are skilled in communicating support and are the same individuals as reviewees.

Some Key Definitions

Outcomes

Outcomes are both goals or objectives and the actual results. An *expected outcome* is a goal or objective. Expected outcomes are declared before the implementation of the intervention. An *actual outcome*, also called an observed outcome or, simply, an outcome, is the effect of the intervention (the service, program, or treatment) on the individuals who receive the intervention. The expected outcome for a child or family is what is determined to be best for the child. In child welfare, for example, an expected outcome may be a child's reunification with his or her parents after placement in foster care. If reunification is not an option, the child is moved quickly to a safe, stable alternate permanent family through guardianship or adoption.

Outcomes also may be neutral or negative. An example of a neutral outcome is that the child, as an emancipated minor, successfully lives independently without the help of the child welfare agency. Another neutral outcome is the transfer of the child's case from one worker or agency to another.

A negative outcome might be that the child's case remains open over too long a time period. Another negative outcome is that the child returns to foster care within six months of discharge because of adoption disruption, failure of the guardianship arrangement, or subsequent reports of the child's maltreatment by the parents.

Process

In this book, *process* refers to a series of interconnected activities performed with or in behalf of the child to achieve positive final outcomes for the child and, usually, the child's family. Process includes completing assessments and case plans and ensuring that children and families receive services as planned.

Variables

Several types of variables are related to outcomes. Variables may be dependent, independent, intervening, or cross-cutting.

The dependent variable represents the phenomenon the intervention seeks to improve. It is the value or condition that is assessed to determine whether improvement has occurred. Any number of dependent variables can be conceptualized for review, evaluation, or study. Examples of dependent variables for purposes of quality assurance are the completion of the service plan and the adequacy of completed plans. An example of a dependent variable in child welfare research is the likelihood that a parent will mistreat his or her child.

The independent variable is the intervention that is intended to bring about a desirable change improvement in the dependent variable. Program services, worker contacts with parents, children's counseling sessions, and other types of interventions can be independent variables in quality assurance projects. If, for example, the dependent variable is the likelihood of a parent's maltreatment of his or her child, the independent variables might be participation in counseling and parent training classes. If the dependent variable is caseworkers' completion of service plans, direct supervision may be an independent variable.

The language used with dependent and independent variables avoids the term *cause*. Causation is not connoted between the independent and dependent variables.

An intervening, or control, variable represents a phenomenon other than the intervention, which is known to have or is suspected of having a definite role in how the dependent variable improves or worsens. An intervening variable diverts the intended effect of an independent variable. For example, if caseworkers' completion of service plans is the dependent variable, intervening variables might be caseworkers' years of schooling, type of degree (social work or non–social work), and years of child welfare experience. If parental maltreatment is the dependent variable, the effectiveness of match of the assigned caseworker and the child, parent, or family might be an intervening variable.

A cross-cutting variable is a characteristic that can be used in analysis to help identify differences among important subgroups. As an example, in a case record review, a quality researcher learns that 30% of a randomly selected sample of case records have incomplete documentation. The researcher might consider which teams completed which records as a cross-cutting variable. The professional might find that all of the incomplete records were the responsibility of two teams, both of which were under the direction of the same administrator. By bringing a cross-cutting variable into the analysis, attention can be specifically targeted where needed.

The Phases, Steps, and Tasks of the Quality Assurance Process

Not only are all phases in the quality assurance plan important, but the phases must be completed in order. Each phase builds on the previous phase. Each of the quality assurance phases consists of specific steps, each of which is based, at least in part, on the outcomes of previous steps. In this book, the phases in the quality assurance plan and the steps in each phase are presented in the order in which they should be performed. It is important that each step be completed accurately, or the entire project can be harmed.

On the other hand, sometimes completion of a later step will clarify an earlier step and lead to the conclusion that the plan specified for a previous step can be improved. In most cases, the appropriate response in such situations is to return to the earlier step, modify the plan, and redo the work that has been done. If changing the plan or redoing the work is impractical, the alternative is to develop a strategy to work around the actual or potential threats. Flexibility is key to the successful performance of the quality assurance process.

This guidebook covers 19 tasks associated with preparation for quality assurance projects (see Table 3). In step-by-step format, it explains all tasks that the quality professional or evaluator must complete before starting to collect information. Such preparation minimizes problems and delays later, when the emphasis is on deadlines and the production of meaningful reports of the findings.

It is important to note that a quality assurance plan is seldom executed precisely as laid out, which explains the importance of project management skills. Paradoxically, this is one of the most important reasons for having a project plan. New and unanticipated issues inevitably arise as a project is implemented, and sometimes, they are sufficiently important to require mini-projects on their own. Without the focus and refocus of a plan, the effort can go off track, never arriving at the original destination or the expected outcomes.

Use of the Guidebook

Application of Steps

In each part of this guidebook, a group of steps for completing Phase A (preparing to review, evaluate, and study) is presented. The guidebook organizes the steps as follows: Part I, Steps 1 through 7: Defining the Issues and Participants; Part II, Steps 8 through 13: Designing the Review and Evaluation; and Part III, Steps 14 through 19: Organizing the Review and Evaluation. Each step specifies a set of tasks to be performed. Each section presents the reader with opportunities to apply the skills presented in the current and previous steps.

Binder Tasks

Most steps have activities entitled "Binder Tasks." When undertaking any review or evaluation, readers are encouraged to compile a binder. The project binder is a repository for the project history and for documentation of every decision and action for the project. The information presented for each step details the specific activities to be recorded in the binder for that step.

The Stellar Agency

To help readers understand concepts, enhance their skills, and apply the information provided, the guidebook refers to a hypothetical organization called Stellar Agency. The Stellar

TABLE 3: Step-by-Step Outline of Phase A: Preparing to Review, Evaluate, and Study

Defining the Issues and the Participants

1. Describe the issues, purposes, and goals of the evaluation and review.
2. Write definitions for all phenomena, the study topics, and related topics.
3. Identify all relevant customers.
4. Identify the relevant expected outcomes.
5. Identify the populations to be reviewed.
6. Define the users of the review findings and their information needs.
7. Choose the methods, techniques, and tools for the review.

Designing the Review and Evaluation

8. Choose an evaluation type.
9. Develop, modify, and select the instruments and data collection forms.
10. Develop materials to train interviewers.
11. Design the samples.
12. Draft the preliminary data analysis plan.
13. Design the layout for the computer database and other computerized files.

Organizing the Review and Evaluation

14. Identify the reviewers and schedule the review.
15. Select the sample.
16. Set up project files.
17. Make sample and reviewer assignments.
18. Deliver review packets.
19. Contact site staff to ensure that records and files are available.

Agency is a private, nonprofit, social services organization with 300 employees who work at four sites around the state. The sites are centrally located within the organization's four service regions. The principal clients are children and families, but the agency also serves individuals, groups, and other organizations. Each client group (or external customer) has certain requirements and expectations regarding the products and services that Stellar Agency provides.

Personnel

Three-quarters (225) of the employees are direct service workers. The remaining 25% of personnel are in support, administration, and management positions. The distribution of employees and clients is shown in Table 4.

Leadership

The president of Stellar Agency is Nigel Garcia. He has a master's degree in social work, has worked for Stellar Agency for eight years, and has been its president for the past four years. On assuming stewardship of the Stellar Agency, he and his leadership team immediately began to shape the organization's structure and culture to achieve expected organizational outcomes. The managers of the agency's five major divisions (direct service operations, human resources, finance and budget, information technology, and knowledge management) organized their divisions to maximize alignments in processes, carry out cultural changes, and support achievement of the expected outcomes.

The largest division, direct service operations, is led by Guillermo Szasz, a licensed clinical social worker with several years' experience in planning and managing. Guillermo designs the programs and implementation strategies. His priority is to ensure that the work processes and service flows work in such a way that the needs and expectations of the agency's social service clients are met.

Terry Bisiriyu manages the human resources division. He has managed employees in a variety of settings. He develops jobs within the organization that respond to organizational needs and are consistent with work processes already in place. He also ensures that supervision and performance appraisals support the achievement of expected outcomes and that conditions at all work sites facilitate positive job performance.

Song Lee Bernstein, a certified public accountant and MBA who has spent her career working in public and private social services organizations, is in charge of Stellar Agency's finance and budget division. This division plans, manages, and tracks the budget and oversees all the organization's financial dealings, including payroll and timekeeping. Song Lee's division includes the support services that ensure that all needed equipment (telephones and voicemail, computers, copiers, and fax machines) operates effectively.

Shaniqua Grabowsky, a certified information technology (IT) professional, oversees the IT division. Shaniqua engineers the organization's computer connections and software development. She is currently testing the integration of case planning and management software with the organization's existing system. In addition, Shaniqua, who designed and led implementation of Stellar Agency's quality system three years ago, comanages the agency's quality activities with Jean-Claude Washington. She is in the process of becoming a certified quality professional.

TABLE 4: Make-up of Stellar Agency					
SITE	A	B	C	D	TOTAL
Employees	53	96	80	71	300
Direct service workers	40	72	60	53	225
Individual clients (all ages)	600	1,080	900	795	3,375
Client families (average size = 4)	150	216	180	159	705
Groups (unrelated people)	75	155	130	100	460
Organizations	2	20	25	18	65

Jean-Claude Washington is at the helm of Stellar Agency's newest division, knowledge management. Many of knowledge management's functions were originally part of the IT division. When Nigel assumed the position of president, however, he foresaw a range of possibilities in knowledge management and created the new division.

The knowledge management division produces literature reviews in areas pertinent to direct services, including clinical methods; human resources; management; local, regional, and national policies and laws affecting social services, service delivery, and organizations that deliver services; information technology and software; and Internet-based tools. The knowledge management division integrates new findings and feasible theories with the existing knowledge base, so that the effects of new information on what is already known can be anticipated. Jean-Claude oversees a product based on this integration, a newsletter, Knowledge Bytes, which is distributed primarily in electronic format, with a small number circulated in hard copy. Originally, Knowledge Bytes was produced exclusively for Stellar Agency's employees. As word spread about the newsletter and its usefulness, however, other social services organizations asked to receive copies. Now, subscriptions are a source of revenue for Stellar Agency. Jean-Claude also is responsible for the agency's library and comanages the organization's quality activities with Shaniqua. He is pursuing certification as a quality professional.

Each of the five divisions has a leader-manager at the four sites. The five leader-managers at each site comprise the site leadership committees. There are five site leadership committees. Members of the site committees rotate as participants of the organization's overall leadership team.

Quality

The central leadership and the leadership committees at each site comprise Stellar Agency's organization-wide leadership council. The leadership council sets the vision and mission and plans how each will be realized. The plan is based on the framework of the quality system with the three main functions of quality management, quality assurance, and quality improvement. Every two years, the leadership council plans and implements an organization-wide review to (a) assess the degree to which organizational and divisional expected outcomes have been achieved, (b) learn whether the stated expected outcomes continue to reflect the requirements and expectations of external customers and other stakeholders, and (c) identify any specific areas that require special attention over the next two years through interventions, reviews, evaluations, or other studies.

PART I

Defining the Issues and Participants

This section describes and applies the first seven steps for preparing to review, evaluate, and study an intervention or program. These steps are:

1. Describe the issues, purposes, and goals of the evaluation and review.

2. Write definitions for all phenomena, the study, and related topics.

3. Identify all relevant customers and providers for the processes and outcomes being studied.

4. Identify relevant expected outcomes for what will be studied or reviewed.

5. Identify the population to be reviewed, unit of review and analysis, and data sources for population measures.

6. Determine the users of the review findings and their information needs.

7. Identify likely data methods and project tools.

This discussion uses a specific format to describe each of these steps, as follows:

- Tasks: Specific actions and supporting information about the actions, such as the reasons for them, how to do them, and their connections to previous or later steps.

- Definitions: Definitions of terms that might be misunderstood or unknown to the reader.

- Discussion: More information or background on the tasks.

- More information: Additional information and the rationale for the step.

- Leaders: Who acts in the primary (and secondary) quality role for the tasks: developer-manager, reviewer, or data analyst.

- Skills: Needed skills from the list of quality skills (see Appendix A).

- Barriers: Any impediments and their effects.

- Binder tasks: Any specific items to be recorded in a personal workbook for the project.

- Application of tasks: Illustration of how the step applies to Stellar Agency.

- Practice questions: A hypothetical or actual quality situation or questions to consider.

- Examples: An application of the concept or skill presented in the step with an explanation of how the step is connected with other steps, when appropriate.

- Additional readings and references.

STEP 1:

Describe the Issues, Purposes, and Goals of the Evaluation and Review

Tasks

- Specify and define the study topics. Most reviews will have two study topics: one topic related to processes (Did the internal or other providers do what they were supposed to do?) and one related to outcomes (Were the clients' or other customers' needs, as defined by the process, met?).

- Write out all that you know or can gather about the topics. Check the literature to learn how the study topics might relate to one another and how they might be related to other aspects of child welfare policy or practice.

- Focus on the purpose of the review. What is it that you need to know after completing this review? What decisions do you need to make? What types of actions will you take based on the review findings?

- Write out the questions you want to answer once you have obtained the results.

Definition

The *study topic* is an aspect of the situation about which you need to have more information before you can reach a conclusion or make a decision. Examples of study topics are: frequency of sibling visitation, adequacy of documentation in case records, and the proportion of all case closures with each defined permanency outcome.

Discussion

The purpose of the review determines the sources and types of data required. There are many review purposes: to measure compliance with a policy or procedure; to determine geographical differences; to assess reasons for noncompliance; to assess the completeness of documentation; to learn staff's views and opinions concerning procedures; and to determine the fit between assessment results, service plan, and service delivery. For example, a review to determine the level of compliance with policies is a study to determine whether the agency assigns Spanish-speaking families workers who can communicate with them effectively. If the reviewer already knows that only a small percentage of Spanish-speaking parents are assigned bilingual workers, the review may identify the reasons for this situation.

More Information

Defining the goal of the review is a problem-solving process. It resembles the actions involved in determining with a client the goals on which the client and caseworker will work by making decisions regarding what must be accomplished to achieve success. Possible data sources for defining the goal are

- Case records,

- Information in databases,

- Face-to-face interviews,

- Surveys,

- Monthly or quarterly reports,

- Direct observations, and

- Focus or discussion groups.

Leaders

Developer-Manager and Reviewer. The developer-manager has the global perspective—the big picture—necessary for assessing how problems might be connected with other organizational practice and policy issues. The reviewer, through working with direct service workers and supervisors, is likely to be aware of circumstances that might not have come to the attention of managers and administrators. The leadership team of the developer-manager and reviewer should consult other quality improvers, as necessary, to develop the concepts and describe the purpose and goals of the evaluation.

Skills

- Administering the review process.

- Employing evaluation methods.

- Employing problem-solving methods.

Barriers

- Not understanding how work gets completed, what the objectives are, or the steps to achieving outcomes.

- Mistaking an approach for a goal or objective and not understanding the implications of the actions that the results require.

- Failing to check with others or the literature for additional information on the issues.

- Denying that the issues are important or that the problem exists.

- Misunderstanding the connections among processes.

Binder Tasks

Write detailed descriptions of the problems and any related issues from practices or policies. Describe how you believe the identified issues are related to the problems. List the likely data sources at this time (additions and changes can be made later). Insert the lists in the binder.

Application of Tasks

The leadership council of Stellar Agency was concerned about several issues. They wondered what caused direct service caseworkers' failures to complete their work in a timely manner. For instance, the council had heard from some supervisors and referring agencies that caseworkers had lags in turnaround for required admission paperwork, missed or late face-to-face visits with clients, and delays in obtaining support services for consumers with multiple needs. Based on their own experiences in the field, the council had some theories about the delays' causes. Some thought the failures were the fault of the caseworkers themselves; others thought they were attributable to supervisors or managers who did not provide concrete resources or guidance. The leadership council also specu-lated that the workers might not understand the agency's policies and procedures on docu-mentation, service frequency, and support service eligibility. The council organized the issues into a few central review questions:

- What is the average time that workers in direct service operations wait for the resources they need to be fully prepared to work with a specific client?

- How knowledgeable are employees, especially caseworkers in direct service opera-tions, about the expected times for completing each phase in service delivery?

- Do clients view their services as being received in a timely manner?

- What are clients' views regarding issues that might affect timeliness of direct ser-vice workers' providing services to them?

Practice Questions

1-1. What are some of the other main questions, suggested by the situation de-scribed, that could be asked to set direction of the review?

1-2. What additional questions will the developer-manager and reviewer ask to sharpen the focus of the direction set by the main questions?

Additional Readings and References

Babbie, E. R. (1998). *The basics of social research.* Florence, KY: Thompson Learning.

Sluyter, G. V. (1998). *Improving organizational performance: A practical guidebook for the human services field.* Thousand Oaks, CA: Sage.

Step 2:

Write Definitions for All Phenomena, the Study, and Related Topics

Tasks

- Identify the dependent and independent variables for the review. Which phenomenon do you want or expect to change over time? For example, the dependent variable could be *status at the time of case closure*. The independent variable could be an aspect of the service delivery process, with the goal being to maximize the proportion of children's moves that are to permanent, stable, and nurturing homes. Breaking the dependent and independent variables into their component parts is essential to capturing all the variables' aspects in their definitions and in measurement instruments. For example, the service delivery process can be broken into components, such as length of service, number of worker-client contacts, and resources used by workers.

- Develop definitions for the dependent and independent variables. After breaking the variables into their components, write precisely what each component means in this review or evaluation.

Definitions

The *dependent variable* is the condition that programs and interventions try to affect. Examples for child welfare agencies are the ability of parents with certain characteristics to keep their children safe, and the ability of foster parents with certain characteristics to provide stable environments for children placed in their homes. In a study of the effectiveness of this guidebook, the dependent variable could be the level of quality assurance knowledge and skills that guidebook users have.

Independent variables are programs and other interventions that are used to achieve desirable outcomes. For example, evaluation and treatment for profound depression and parenting training are independent concepts.

An *intervening factor or variable* is anything that enhances the effect of the programs and interventions or that keeps the interventions from working as expected. For example, a possible intervening factor could be the number of times that the parent did not participate in treatment due to illness or lack of child care, or the times that the program leader could not follow through due to absence. There also can be external factors. An unsafe neighborhood might prevent a parent from keeping his or her child completely safe. On the other hand, the parent's social support network can enhance the effectiveness of the parent's participation (unless building such a network is part of what parents are helped to do).

A *cross-cutting variable* is a "filter" used to keep the dependent variable separated into classifications or categories that should not be grouped together. Phenomena used as cross-cutters depend on the specifics of a study. For example, Stellar Agency will use geographic site as a cross-cutting variable in all agency-wide reviews, evaluations, or other studies, because they must know which sites have problems and which improvements to make at each of the four sites.

Discussion

Writing a definition is almost a surgical process. Every word adds to the meaning, and unnecessary words muddle meaning. As a result, it is vital that only words that clarify meaning be used. Evaluate the need for each word before adding it the definition. Scrutinize what has been written to detect words that might reflect or cause bias, and change any such words.

Seek review of and feedback on the definition from others who might view the topic differently. Continue getting feedback until you are comfortable that what is written conveys your intended meaning to most readers.

Leaders

Developer-Manager and Reviewer. The developer-manager should have a background that includes research methods. Because the concepts of dependent and independent variables are fundamental to the study of research, the reviewer should be familiar with these concepts, although, in some cases, he or she may need clarification from the developer-manager.

Skills

- Employing research methods and process.

- Developing and designing case reading and other data collection forms.

Barriers

- Incorrectly believing that a written definition is objective when, in fact, it is biased.

- Failuring to unpack complex concepts so that the essential, indivisible components of the concepts can be defined.

- Not taking the time and effort to avoid words that could make the variable definitions less specific or that exclude concepts intended to be covered. Every single word in the definition affects meaning; if any word is changed, the definition changes.

Binder Task

Write the definitions for dependent, independent, intervening, and cross-cutting variables. Keep them in the project book, and refer to them frequently. Some definitions may change as measures for the defined concepts are developed.

Application of Tasks

Stellar Agency identified the following variables for the reviews that it planned:

- Dependent variables:

  - Mean wait time for direct service workers to obtain required, effective access to various services or equipment (telephone, receipt of messages, copiers or copy service, supplies, and personnel).

  - Mean wait time for direct service workers to obtain required, effective input, such as supervision, general advice, or answers to specific questions, from supervisors, managers, or colleagues, or for direct service workers to obtain the resources they need to be fully prepared to work with a specific client. *Required* connotes that work cannot continue without the service, equipment, input, or resources; effective connotes that the input received is good enough to allow work to continue.

- Independent variables:

  – Types and adequacy of service and equipment: available telephones, effective message receipt, available copier or copy service, on-hand supplies, number of full-time-equivalent staff, and number of vacancies that were to be filled.

  – Types and adequacy of input: frequency and adequacy of regular supervision from direct supervisor, response time, and adequacy of answers to questions asked of supervisors and managers. *Adequacy* refers to whether the service, equipment, or input satisfies the direct service workers' needs.

Practice Questions

2-1. Identify one independent variable that might be included in Stellar Agency's current project.

2-2. What is Stellar Agency's cross-cutting variable?

STEP 3:

Identify All Relevant Customers and Providers for the Processes and Outcomes Being Studied

Tasks

- Identify all relevant internal and external customers, customer providers (individuals or work units most responsible for meeting customers' needs), and other relevant providers (employees) for the customers' needs. Internal customers are just as important in the processes and outcomes as service consumers and other stakeholders.

- Specify all customers' needs. To ensure that everything you want to review or study is included, compare the specified requirements to the outcomes identified in the previous step.

- Specify which individual or work unit is or should be responsible for ensuring that customers' needs are met.

- Prioritize the list of customers and specify high-priority providers. Virtually everyone in or associated with an organization is a customer of some type.

Definitions

A *stakeholder* is any individual, group, or other entity with an interest in whether an organization, department, work unit, or individual uses a particular process or achieves a particular outcome. As an example, the U.S. Congress may be viewed as a stakeholder in the number of adoptions of children in foster care that are finalized each year. The following are examples of stakeholders.

A *customer* is any individual or entity, internal or external, that uses products or services produced or delivered by the organization. *External customers* are not part of the organization's workforce. The primary external customers are people to whom the company delivers products or services. For example, the primary external customers of a child and family agency, such as Stellar Agency, are clients, which include children, families, and groups. The primary external customers of a restaurant are diners, and the primary external customers of an ice cream shop are the purchasers of ice cream cones.

Internal customers may be a new concept for employees. Thinking of workforce members as customers, however, helps heighten awareness of the importance of ensuring that front-line employees have the resources they need to meet the requirements or expectations of the primary external customers. The primary internal customers of Stellar Agency are direct service workers.

Examples of customers are

- clients, such as children, families, individuals, groups, or organizations;

- foster parents;

- any staff member who, in doing a job, requires something from another staff person;

- boards of directors, trustees, and advisory boards;

- state legislatures; and

- agencies with contractual relationships.

A *provider* is any individual or entity from whom a customer receives something needed to complete necessary tasks or achieve an outcome. Every customer depends on at least one provider, and most customers depend on multiple providers. Almost every individual or work unit is a provider in some way. Providers are defined by who expects or requires something that the individual or entity can supply.

At Stellar Agency, direct service workers are providers to the children and families served by the agency. Supervisors are providers to the direct service workers. The personnel who maintain up-to-date data in Stellar Agency's information systems are providers to the direct service workers and their supervisors, the managers who guide and support the supervisors, the quality professionals who examine the work being done, and the clerical staff who process and file paperwork.

Leaders

Developer-Manager and Reviewer

Skills

- Employing research methods and process.

- Employing problem-solving methods.

 Binder Task

List all employees and other staff who depend on the results of the process to do their jobs. Specify what each individual needs and for what purpose the results will be used. You may have to use persistence to obtain the answers to complete this task.

Application of Tasks

Stellar Agency's most important customers are its primary external customers, that is, children, families, and groups. Next are the organizations that receive products from the knowledge management division. The agency's most important internal customers are the individuals who are the primary providers to direct service clients—the direct service workers. Staff in the knowledge management division are also important internal customers, because they produce something that is used by external customers.

In one way or another, employees in all divisions have a stake in the findings of the review being planned. Practically speaking, human resources, finance and budget, information technology, and knowledge management are providers to direct service staff. Each division's purpose is to ensure that direct service staff have what they need to function effectively. Within the direct service division, staff in positions other than direct service also ensure that direct service staff have what they need to perform their jobs.

There are several entities with a stake in the results of Stellar Agency's biannual review and the more specific projects being planned, such as the state agencies with which Stellar Agency has contracts to serve vulnerable populations; other funding sources; state lawmakers; certain federal agencies; the businesses from which Stellar Agency buys equipment, supplies, and services; and the universities that provide training and staff development workshops. Although the findings of the biannual reviews are routinely shared with many external stakeholders, most results and findings of the current project will be for internal consumption only.

Practice Questions

3-1. Who are some of Stellar Agency's other internal stakeholders? Explain.

3-2. Who are some of Stellar Agency's likely external stakeholders? Explain.

STEP 4:

Identify Relevant Expected Outcomes for What Will Be Studied or Reviewed

Tasks

- Restate each relevant customer's requirements and expectations in terms of outcomes to be achieved. These outcomes will be the expected outcomes for the issues being studied in this evaluation.

- List all the processes and the most important tasks necessary for achieving the expected outcomes.

- Determine where measurements of the processes and outcomes might be obtained.

- Review Step 3 for the descriptions of customers and providers.

Definitions

A *requirement* is a product or service, provided within a specific time frame, delivered in a specific format on the basis of a binding rule of the organization.

An *expectation* is a product or service in the form that a customer or stakeholder desires. An expectation about a particular product might be the same as or different from the requirement.

An *expected outcome* is something that is sought. Expected outcomes are desired results that are declared before interventions are implemented. Expected outcomes are sometimes called *goals* or *objectives*.

Leaders

Developer-Manager and Reviewer

Skills

- Employing research methods and process.

- Performing a stakeholder analysis.

 Binder Task

Write statements about the expected outcomes and the processes used for achieving outcomes. Describe in detail how you think the processes and outcomes are connected. You will use this information later, particularly when drafting the data analysis plan and interpreting the findings from the analyses.

Application of Tasks

Expected outcomes are based on the requirements and expectations of the customers pertinent to the specific review, evaluation, or study. Assume that the Developer-Manager in charge of the project has recently checked with the pertinent customers to verify their requirements and expectations. Table 5 shows some of stakeholders of the Stellar Agency, their requirements and their expectations.

Practice Questions

4-1. Identify at least one additional expectation for each of the three specified customers.

Example

Performing a Stakeholder Analysis

A stakeholder analysis is used to identify or clarify stakeholders, their specified and unspecified requirements, and their expectations. It provides a means for specifying outcomes.

See Table 6 for an example of a basic stakeholder analysis worksheet that can be tailored to fit any organization.

1. Make a table with at least six columns and use the following headings:

   a. **Stakeholder**. An individual, group, organization, or some other entity with something to gain or lose due to the organization's existence, functioning, or products and services.

   b. **Type**. Information in this column describes how the stakeholder is related to the organization, such as external or internal customer, primary customer or other type of customer, and other information.

   c. **Requirements**. This column states what the organization is committed to provide to the stakeholder, such as a service or product, technical assistance, a report, or an agreement.

   d. **Expectations**. This column indicates what the stakeholder thinks will be provided, which could be different from what the organization is required to provide. Information from the stakeholder may be needed to complete this column.

   e. **Frequency and delivery**. This column states how often a service is provided—daily, weekly, biannually, or other. It also indicates how a service, product, or other deliverable is provided.

   f. **Other**. This column captures any aspect of the relationship with the stakeholder that has not been documented.

2. Look for known requirements, examining the organization's mission statement as a start. Consult the proposal, contract, charter, bylaws, enabling legislation, or any other official documents to verify the intended primary and other beneficiaries of the program and what they are supposed to receive from the organization. Be diligent in this search. In addition, study recent communications from stakeholders, which could yield requirements and expectations.

3. After looking through the documents, consult known stakeholders to obtain a clear understanding of their expectations. Use focus or discussion groups—not surveys—to ensure that you fully understand what stakeholders expect. Fill in the appropriate information in the table.

TABLE 5: Customers, Their Requirements, and Their Expectations

Customer	Requirements	Expectation
Direct service client	Timely interventions and receipt of services from direct service workers (the provider of this requirement)	Stellar Agency staff's awareness and understanding of the clients' circumstances and viewpoint
	Service continuity	
	Inclusion by direct service workers in service planning and implementation decisions	
	Current assessment and service plan used by the direct service worker reflects client's actual circumstances	
Direct service worker	Adequate resources and input from supervisors or other provider of this requirement received in a timely manner	Adequate resources and input, received in a timely manner
		Moral support
		Supervisory and management awareness of the direct service viewpoint

TABLE 6: Stakeholder Analysis Worksheet					
Stakeholder	Type	Requirement	Expectation	Frequency and Delivery	Other
Child welfare organizations	Primary, external	Accurate and current knowledge on effective practice	Accurate, current information delivered in various modes	Quarterly, with specials projects as needed	
Children and families	External, nonprimary		Assigned workers with current knowledge	Undetermined	
State child welfare agency		Reports on outcomes	Reports on outcomes	Quarterly, written reports and e-mail	
Day care organizations with education components					

Step 5:

Identify the Population to Be Reviewed, the Unit of Review and Analysis, and Data Sources for Population Measures

Tasks

- Determine the make-up of your population. What people or things will you study? Because the group you will examine often is part of a larger group, the essence of this task is the selection of the characteristics that should be excluded from the study. The definition of the population must allow you to decide whether each characteristic of the group should be included in the study.

- Identify the people or things you want to make statements about, based on the results. Review the purpose established for the review and evaluation (see Step 1). You may want to make statements about people (customers, providers, or individuals responsible for certain requirements), processes (events, incidents, or occurrences), or other phenomena.

- If you will be sampling records from an existing data file, request and keep a copy of the format of the file that is provided and a complete description of the field types, field widths, and the position in the file of each field. You will use this information in completing some of the later steps.

- If any of populations in the study involve clients, contact the agency's Institutional Review Board (IRB) and start the process for IRB approval of the aspects of the review that involve clients. Some IRBs require that data collection forms and instruments be submitted with the application for approval. If the IRB requires this information, you will need to complete the tasks of Step 9 before beginning the IRB approval process. Although obtaining IRB approval is not technically part of the main procedures for this step, the failure to address the IRB process early can delay your project.

Definitions

The *population* (or *universe*) is the larger group that you will study and about which you will draw conclusions and, perhaps, make recommendations. The population's characteristics mirror the characteristics represented in the dependent variable including any that should be excluded from the definition. Sometimes this means specifically excluding certain characteristics from the definition. As an example, if the study will examine how intake procedures are handled, the population definition must clarify which cases are not parts of the population, such as cases that have been open for more than a designated period of time.

As an example, if the study will examine how intake procedures are handled, the population definition must clarify which cases are not part of the population, such as cases that have been open for more than a designated period of time. Given that definition, there is no chance that such cases will be sampled, that is, these cases have a zero probability of selection. For example, if the developer-manager decides to limit the review to children in out-of-home care, that exclusion must be made explicit. Similarly, if the study is to examine only children who are placed with foster families, and not children in a residential placement, that exclusion must be stated.

The *unit of analysis* is the specific, discrete thing being studied. The unit of analysis is tied to the population, because population elements are usually the main (and, sometimes, the only) unit of analysis. For example, in the intake study example above, the unit of analysis is the individual case of each child. Larger units may be constructed from the main unit by aggregating cases. Sometimes cases are aggregated to take advantage of data that are measured only at a higher level of abstraction, such as families, rather than individuals. Any conclusions drawn from analyses with aggregated units and data at higher levels of abstraction, however, cannot be applied to the individuals that comprise the aggregates. Doing so would be a violation of the *ecological fallacy*, in which any conclusions drawn from analyses with aggregated units and data at higher levels of abstraction cannot be applied to the individuals that comprise the aggregates.

Field type indicates the form in which the data are kept, such as numeric or string.

Field width indicates how many spaces are allocated for the data, such as two spaces to record a client's age.

Position indicates where the field or data can be found in the existing data file, such as first, second, and so forth.

More Information

Possible populations for child welfare reviews and studies include

- all children in care for 30 months or more;

- all families in the child welfare system as the result of more than one substantiated report of suspected child maltreatment;

- all foster parent grievances or requests for appeals filed in the past six months;

- all children in care who have siblings who they have not visited with in the past six months; and

- all children in care who have visited with a sibling within the past 12 months.

Leaders

Developer-Manager and Reviewer

Skills

- Selecting characteristics to be included and excluded in the population definition.

- Designing and selecting samples for quality reviews.

- Speaking effectively at meetings, during group processes, and in one-on-one exchanges.

 ## Barriers

A poorly defined independent variable. The study population is the total of whatever or whoever has the characteristics of the dependent variable. If the dependent variable is not defined will, it will be difficult to correctly specify the population.

 ## Binder Task

Describe the population, the unit of analysis, and the data sources.

Application of Tasks

Stellar Agency defined several stakeholder populations for purposes of this project:

- direct service clients served under a contract with a state agency—cases with individual children as the primary client, families as the primary client, individual adolescents as the primary client; and individuals 18 and older as the primary client;

- direct service clients not served under contract with a state agency, using the same specifications as above;

- clients who receive services, such as counseling or directed discussion, in groups;

- organizations that receive Stellar Agency's newsletter, *Knowledge Bytes*;

- direct service workers at each of the four sites;

- supervisors of direct service workers at each site;

- local site leadership committee members at each site; and

- clerical employees working in direct service operations at each site.

The clients and staff receive the results of the staff's efforts. Their condition tells the effects of the staff's work.

Practice Question

5-1. What will be the likely unit of analysis for data from each population specified in Stellar Agency's list of stakeholders?

Determine the Users of the Review Findings and Their Information Needs

Tasks

- Identify the report consumers so that the evaluator can verify their information requirements and collect any additional data needed to meet those requirements. By understanding consumers' needs, the organization can enhance its understanding of the results. Two types of report consumers exist: (1) individuals likely to absorb the information and apply the findings to make improvements on the job, and (2) individuals interested in and likely to read the report, but unlikely to apply the findings to work processes and routines. Both types of consumers should receive reports. The first type is likely to want to receive reports with helpful details quickly; the second type is generally satisfied with report summaries.

- Talk to the report consumers to determine what type of report is best for them. Detailed reports do not have to be wordy. Graphics, charts, and tables can communicate details or overviews. Some report consumers prefer text; others prefer graphs. Find out which form consumers prefer, and make the effort to produce each report type. Similarly, some report consumers want hard copies and others prefer electronic formats. Begin recording report preferences to avoid delays in report preparation when the results are ready.

- Another issue regarding reports is the frequency with which they are issued. One of the biggest mistakes made in disseminating reports is releasing them at the same frequency to all users. Some report users will not be able to absorb the information very quickly and may, as a result, begin to ignore all reports they receive. The organization's leadership needs frequent (usually daily) reports to steer the agency's course. Most direct service workers are satisfied with monthly, bimonthly, or quarterly reports.

Leaders

Developer-Manager and Reviewer

Skill

- Speaking effectively at meetings, during group processes, and in one-on-one exchanges.

 Binder Task

Record the report preferences of each person who will receive findings.

Application of Tasks

The findings of this study will not be circulated externally. Stellar Agency employees will receive those sections of the study findings that apply to them. Clients who participate in the study will receive a summary of the findings that pertain to the information that they provided. This aspect of the dissemination plan may be revised in a later phase of the project, but it is important to consider it at this stage, because information requirements may affect data sources or collection methods. Table 7 outlines the groups that will receive findings and the types of findings that they will receive.

Practice Question

6-1. Stellar Agency could post the findings on its Internet site, but it has chosen not to do so. Why?

TABLE 7: Recipients of Information and Types of Information to Be Provided	
Recipient	**Information Type**
Leadership council member	The executive summary and detailed results and recommendations
	Print and e-mail versions
	Formal presentation by Saniqua and Jean-Claude, who comanage quality activities
Employee	Summary of most findings, and details of sections that affect them and their clients
	E-mail version
	Presentation by local leaders, with Shaniqua's and Jean-Claude's support and assistance, presented in person and through distance delivery media, repeated several times to ensure that all employees can attend the presentation
Client	Summary of findings on those items to which they responded
	Verbal description of findings during a meeting with the worker
	One-page summary of the information that is shared verbally with them
Knowledge management staff	Summary of most findings and details of sections that affect them and their client organizations
	E-mail version
	Attendance at one of the site presentations for employees

STEP 7:

Identify Likely Data Methods and Project Tools

Tasks

- From the outset, determine where the evaluator will obtain the data and what steps he or she will need to take to select the sample or population elements. Although decisions about data collection methods, analysis techniques, and tools are preliminary at this stage, begin thinking about which methods, techniques, and tools are appropriate and available, and which are impractical or impossible for the planned review. If the optimal data collection methods are not feasible for this project, consider the next best alternative.

- Data collection is part of the variable-measuring process. There are numerous data collection methods (as discussed in the More Information section of this step). Each method requires a feasible strategy for data gathering. Some data collection strategies work better in certain environments than others.

- There also are several analysis techniques. If the project is an evaluation, be aware of the level of measurement of the data and issues of statistical conclusion validity. Step 9 addresses this issue, and Appendix B reviews statistical conclusion and other types of validity. If the project calls for data processing by peers, consult other publications (see Mears, 1995; Zirps, 1997). Remember that you are not making final decisions about methods and techniques at this point, but, instead, are thinking of likely possibilities.

More Information

The most common data collection methods for quality assurance reviews (not necessarily in order of priority) are:

- reading case records or other files to locate information to be measured, and recording measurements on forms;

- observing events to perceive, synthesize, and analyze information to be measured, and recording measurements on forms;

- leading focus groups to discuss topics and provide measures of variables in the process, and organizing the information gathered through discussion into meaningful measurements of variables;

- conducting in-person interviews and mail or telephone surveys (either self- or interviewer-administered) with individuals who have knowledge or have exhibited behaviors (including attitudes), obtaining information for measuring research variables; and

- accessing databases and using data field elements as measurements of variables or to construct variable measurements.

Leaders

Developer-Manager and Reviewer

Skill

- Employing research methods and process. (Appendix C describes the research process.)

Binder Task

Note your choice of data collection methods and the rationale for the selection. Draft the data collection plan, which should include specific steps for locating the data sources and the information.

TABLE 8: Possible Data Collection Strategies

Population	Data Source	Acquisition Strategy
Direct service clients	Individuals and client records	Mailed surveys, in-person or telephone interviews, focus groups
Groups	Group	Surveys, focus group
Employees	Individuals	Web-based or mailed surveys, in-person or telephone interviews
Organization	CEO or designee and records in the knowledge management division of contacts with the organization	In-person interviews

Application of Tasks

The Stellar Agency's project will require several methods. Data acquisition and analysis are founded in research skills and methods. The evaluator has defined multiple populations, and each population requires a strategy specific to its characteristics. Some considerations are whether the population members are in a list, in a computer database, or maintained in some other format; whether the data are available in a database; whether the information must be obtained directly from people, read from paper records, or accessed through other means; and so forth. Specific data acquisition strategies are not being decided, just the likely ones, however, data should be identified now. Table 8 identifies possible data sources.

The Stellar Agency has considered several study questions to guide the review and evaluation process, but decided to use only the following: What is the average time that workers in direct service operations wait for the resources they need to be fully prepared to work with a specific client?

Steps 1–7

The foregoing seven steps lay the groundwork for the project. Before moving ahead, review Steps 1 through 7. Next, update and edit your project binder to make sure that it documents the rationale for the plan and actions clearly. The goal for the binder is to document the project history and become a reference for future projects.

Study the following recap answers to the practice questions for Steps 1 through 7, which illustrate how Stellar Agency could implement each step.

Step 1

1-1. What are some of the other main questions, suggested by the situation described, that could be asked to set direction of the review?

The purpose of asking the questions is to set a direction for the project. The questions will help the leaders frame an inquiry into sources of delays in the process. Asking about length of wait time for resources points to studying whether direct service workers face a specific type of barrier to proceeding with processes. Asking about workers' knowledge suggests investigating their preparedness to do the job. The leadership council might ask a variety of initial main questions and then see if any can be ruled out before proceeding.

The questions the leadership council asks initially represent probable directions of inquiry and depend on the specific circumstances of the organization at that time. For example, if the leadership council thinks it is feasible that direct service workers meet barriers but do not lack resources, the council members might ask questions about adequate staffing (see 1-1A and 1-1B). If they believe that there could be special issues associated with the current client population that slow down progress, the council might pose another question. The council would want to obtain measures from the perspective of clients, because service delivery affects them and their views are important (see 1-1C).

> 1-1A. *What is the average time that workers in direct service operations wait for the resources they need to be fully prepared to work with a specific client?*
>
> 1-1B. *What is the average number of clients assigned to each worker or team of workers?*
>
> 1-1C. *How do clients rate various aspects of the services they receive?*

1-2. What additional questions will the developer-manager and reviewer ask to sharpen the focus of the direction set by the main questions?

Questions 1-1A, 1-1B, and 1-1C (or A, B, and C) represent some of the different possible causes of the increase in wait times.

Suppose that the developer-manager and the data analyst analyzed data in the Stellar Agency management information system and learned that the number of families and the number of individual children assigned to workers is about the same throughout the agency. That would mean that *b* could be eliminated as a question to pursue at this time.

The Stellar Agency's quality assurance division conducts client satisfaction surveys annually. In addition to analyzing each annual survey separately, the developer-manager and data analyst compare findings across years, so that they can detect any trends. The developer-manager examined the findings of the most recent client satisfaction survey and trends of the past few years. He or she learned that the level of client satisfaction was fairly even across the years but dipped slightly in the most recent survey.

The developer-manager and data analyst know that data to answer Question A are not collected routinely. Thus they cannot perform a quick analysis to answer it. Question A will be retained as a question to answer through project activities.

Using historical data from the management information system, the data analyst computed, for the past six months, the average length of time that it takes a direct service worker to complete all work on a client system (i.e., family or child). The developer-manager and data analyst called this *mean wait time*.

The developer-manager, working with the reviewer, posed several questions about mean wait time:

1-2a1. *Does the mean wait time vary among the Stellar Agency's four geographical sites? If mean wait time varies, what accounts for the differences?*

1-2a2. *Are differences in mean wait times random across stages in the process, or does a pattern of longer mean wait times exist at particular stages? If there is a pattern, what causes it?*

The questions are broad and must be broken into all of the components rolled into them. Breaking the questions into specific components allows the developer-manager to develop definitions for what will become the variables.

Step 2

2-1. Identify one independent variable that might be included in the Stellar Agency's current project.

Complexity of circumstances of the clients assigned to direct service workers is one possibility. If complexity of circumstances (or severity of problems) results in more time required to work with clients and lengthening the time these clients remain in the system, and complexity of circumstances is not distributed evenly among all direct service workers at all sites, complexity would contribute to longer wait time. Another possible independent variable is severity of clients' problems.

2-2. What is Stellar Agency's cross-cutting variable?

Geographic site

Step 3

3-1. Who are some of Stellar Agency's other internal stakeholders? Explain.

All of the employees who work for the Stellar Agency in capacities other than direct service delivery are internal stakeholders. The are stakeholders because they all have an interest, a stake, in whether direct service delivery to clients—and achievement of all other outcomes—is successful.

3-2. Who are some of Stellar Agency's likely external stakeholders? Explain.

The external stakeholders include all parties with an interest in the Stellar Agency's successful achievement of outcomes. The obvious ones are organizations that contract with the Stellar Agency, such as agencies who refer clients or receive referrals from the Stellar Agency, donors and funding sources, and vendors of supplies. There also are stakeholders that might not readily come to mind: legislatures, taxpayers, and the public in general.

Step 4

4-1. Identify at least one additional expectation for each of the three specified customers.

- Direct service clients: To be treated respectfully by Stellar Agency personnel.

- Direct service workers: To have input in decisions that affect their ability to serve clients effectively.

- Knowledge management employees: For Stellar Agency employees to acquire and share information with their colleagues.

Step 5

5-1. What will be the likely unit of analysis for data from each population specified in Stellar Agency's list of stakeholders?

- Population: Direct service clients (DSCs) served under contract with a state agency
  Unit of Analysis: Individual direct service workers (DSWs)

- Population: DSCs not served under contract with a state agency
  Unit of Analysis: Individual DSWs

- Population: Organizations that receive Knowledge Bytes
  Unit of Analysis: Organizations

- Population: DSWs at each of the four sites
  Unit of Analysis: Individual DSWs

- Population: Supervisors of DSWs at each site
  Unit of Analysis: Individual direct service supervisors

- Population: Local developer-managers at each site
  Unit of Analysis: Individuals who are local leader-managers

- Population: Clerical employees working in direct service operations at each site
  Unit of Analysis: Individuals holding any clerical positions

Step 6

6-1. The Stellar Agency could post the findings on its Internet site, but they have chosen not to do so. Why?

It would be more difficult to control access to findings if they were posted on the Internet. The Stellar Agency wants to make sure that report consumers receive the findings in ways that best suit them.

PART II

Designing the Review and Evaluation

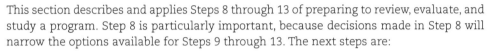

This section describes and applies Steps 8 through 13 of preparing to review, evaluate, and study a program. Step 8 is particularly important, because decisions made in Step 8 will narrow the options available for Steps 9 through 13. The next steps are:

8. Make key decisions

9. Develop or select and modify the data collection forms

10. Develop training materials and train interviewers

11. Design the sample

12. Draft the preliminary data analysis plan

13. Design and document database layouts

STEP 8:

Make Key Decisions

Task

- Based on what you have learned and noted, choose an evaluation type. Look over the notes you made about the purpose and other features of the review being planned. The information in Step 8's discussion may help you make your decision.

Definitions

Validity is the best available approximation to the truth in propositions about the phenomena being studied. *Approximately* and *tentatively* are understood as prefaces to the terms *validity* and *invalidity*, because what is true can never be actually known (Cook & Campbell, 1979). Unlike reliability, researchers have no objective way to test for validity that will produce a coefficient. Rather, the developer-manager must use a logical process to protect the four types of validity (described in Appendix B) and must use rational discussions to discount "competing hypotheses," that is, phenomena other than the intended ones that could account for the obtained results.

Aggregating is summarizing data associated with individual, observed units into groups of data. For example, measures of individual family members' levels of satisfaction can be aggregated by adding individual scores and taking the mean of these scores, averaging the dispersion (the variation in the scores), or applying any number of other procedures based on the purpose of the review.

Discussion

As a preliminary step to selecting the evaluation type, consider the following:

- Refresh your memory on the review and evaluation purpose developed in Step 1.

- Specify what you will need to know after the project has been completed. Review the questions that you developed in Step 2. Look at the questions you set out to answer with the review. Which variables do the review, evaluation, policy, and research questions lead you to include?

- Consider measurement issues. Although measures will not be completed until Step 9, make a note of measurement issues now as a binder task to remind you to address those issues later. Which intervention is being reviewed or evaluated? Can it be validly measured as a whole, or is it a complex combination of many things? Can the intervention be measured directly, or must the evaluator identify proxies to indicate its presence or magnitude? Can you collect measures from one place, or do you need information from multiple sources? Will the data require conversion so that all data are at the same level of aggregation and in like or compatible units across all variables? Developing valid measures is one of the most important components of data integrity.

- What is the intervention supposed to change? What kinds of change and how much change are expected? Think about the practice aspects that could prevent success or support the project. What things could harm validity (see Appendix B)?

- Anticipate the results, findings, and recommendations. Based on the results that you obtain, what actions might you take? What actions might others take if the results indicate deficiencies?

- Could a program or service be discontinued or significantly downsized based on the findings? Might a program or service be expanded? If the answer to either of these questions is yes, take special care with validity and the precision of estimates, including point and interval estimation (see Blalock, 1979).

- Regardless of the effects on programs or services, how precise must your results be? A higher level of precision requires greater attention to detail and very careful implementation of the process. How important is it to maintain a small range of probable error?

- How accessible are the data? Are the data directly available, or must you employ a screener to locate population members and then select from the population members you have identified? Are the data in computer files, paper lists, desk drawers, or in some form that will require manual manipulation? Are the data likely to be plentiful or sparse?

- Do the techniques you considered in Step 7 fit with the rest of the project plan?

After considering these questions, select the review and evaluation type. Options include:

- **Process Evaluations**. These projects, also called formative evaluations and implementation studies, determine whether (a) work is being performed according to defined standards or other expectations, (b) expected linkages actually connect, and (c) the actions taken make sense, given the expected outcomes. Through examining processes, it is possible to determine whether a program is achieving defined outcomes.

- **Process-Outcome Studies**. Process-outcome studies are useful in determining whether the expectations of all customers and other stakeholders are being met or exceeded. They permit a comparison of actual outcomes to expected outcomes. Step 17 details how to make such comparisons.

- **Cost-Benefit or Cost-Efficiency Analysis**. In cost-benefit or cost-efficiency studies, the relationship between project costs and outcomes is examined. Both costs and outcomes are expressed in monetary terms, and results are expressed as costs per unit of outcome achieved, or efficiency.

- **Comprehensive Evaluation**. A comprehensive evaluation examines the conceptualization and design of interventions, the effectiveness of work processes, how closely actual outcomes match the expected outcomes, and the financial implications of these factors.

Leaders

Developer-Manager, Reviewer, and Data Analyst.

Skills

- Employing evaluation methods.
- Employing research methods and process.

Application of Tasks

After considering all of the relevant issues, Jean-Claude and Shaniqua agreed that the project will use two evaluation types: process and process-outcome. One reason for this was that a process study would help determine whether the caseworkers were meeting the requirements in Stellar Agency's policies and procedures, something about which the leadership council had already expressed concern. Another was that a process-outcome study would help determine whether stakeholder expectations, such as those of the referring agencies and the caseworkers themselves, were being met.

Practice Question

8-1. Why did the developer-managers decide not to do a comprehensive evaluation?

Step 9:

Develop or Select and Modify the Data Collection Forms

Tasks

- Decide on the data to be collected, how to obtain the data, and the data collection methods to be used (see Step 9's More Information section).

- Decide the types of forms that will be required and how they will be used. Form designs differ for different methods of data collection (see Examples A through E). All forms should be developed or adapted to maximize ease of the data collection process. Each form's design must match the purpose of the study and circumstances in which the form will be used.

- Determine if anyone developed suitable data collection forms for a past study. Study individual items to determine if they meet the evaluation's needs. Secondary items must reflect operational definitions, which you began developing in Step 3.

- Operationalize the main study variables, specifying question objects (defined below) in the process. The operationalization process is described in More Information.

- Decide on the level at which each question object will likely be measured. Revisit Step 1 and review information on the purpose and goals. The purposes might contain requirements for measuring question objects at particular levels.

- For each question object, draft a question stem. To the extent possible, question stems should be characterized by singleness, that is, question stems should be written as a simple sentence with a single subject, a single verb, and if an adjective or adverb must be used, one modifier. In most instances, it is better to put modifiers in the response categories, not in the body of the stem.

- If an item will be precoded, develop response categories. *Precoding* means establishing codes or categories for question objects on the measurement tool before the researchers collect data, not after. Response categories should be the continuation of the sentence begun in the stem. They must follow the stem in form, and they should be grammatically compatible with the question stem. The stem and response categories are the complete measure.

- If an item will be open-ended, delay the completion of the development of the measure until (a) data collection has been finished, (b) codes reflecting categories of responses have been developed, and (c) responses have been sorted into their respective groups.

- If the evaluation uses a previously developed item, assess its appropriateness against the criteria specified in the operationalization process for developing measures and items in the next section.

- Determine whether the question and item object match the components the evaluators defined during the operationalization process.

- Combine the question stem and the response categories to form the draft item.

- After the items have been written, format them to fit the data source, the data collection methods, and the strategies.

- If the data collection form will be used for the first time in this study, have it reviewed by others. Select at least one each of the following: a professional in the subject area, a person who develops research items, and if appropriate and possible, a layperson who received the service or was involved with the topic and who can offer a nonprofessional view. After revising the draft, pretest it. Modify it according to pretest findings and finalize.

Definitions

The *level of measurement* is the precision with which the attributes of question objects can be described for developing and taking measurements. There are four levels of measurement, the highest level offering the greatest possible precision and flexibility. Some phenomena, however, are inherently not amenable to being measured at higher levels and must be measured at lower levels. The four levels, starting with the lowest, are nominal, ordinal, interval, and ratio (see More Information).

Level of measurement may be confusing to professionals who are new at developing items. Level of measurement is vitally important for accuracy of measures and statistical analysis. The level at which a variable is measured defines the choice of statistical procedures that can be used in the variable's analysis. Statistics appropriate for higher levels of measurement are more powerful than statistics that can be used with variables measured at lower levels. Results from higher level statistical processes are more robust, and their interpretations are more sensitive than results achieved with lower-level statistical processes. Items should be developed at the highest appropriate level of measurement possible.

Operationalize is to define a concept to specify its components. Operationalization is the process of defining and separating into components until all essential parts have been defined and until indicators of question objects are apparent. Before items can be developed (or selected, if the project uses items already used by another reviewer or researcher), the items must be divided into their essential components, or operationalized.

A *question object* is the essence of the phenomenon—a thing, an action, or a thought—for which the evaluator develops measures.

A *question stem* is the body of the item, specifying indicators of the question object.

A *questionnaire item* is an enumerated item on a questionnaire, consisting of the question object, the question stem, and response categories.

More Information

Many studies collect different types of data, from different sources, and in different ways. In fact, researchers design some studies to measure the same question objects in various ways. In these studies, the items they develop to measure the question objects use different words depending on the data source that is being used.

For example, in a study to determine which services caseworkers provide for children, evaluation professionals may wish to obtain information from three sources: children in out-of-home care, the children's caseworkers, and the children's foster parents. The professionals also may be interested in learning whether relevant information about services is documented in the child's case record. To capture such information, three types of forms may be developed: interviewer-administered forms to obtain data from children and foster parents; a self-administered survey to obtain information from caseworkers; and a data form that a review team member uses to collect information from children's case records (see Examples 9-A through 9-D).

Example 9–A

This example shows the process of operationalizing a concept into a funished data collection item.

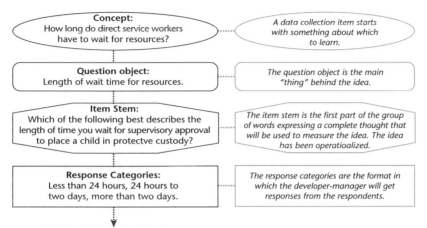

Finished Data Collection Items:
Which of the following best describes the length of time you wait for
supervisory approval to place a child in protecive custody?
Less than 24 hours • 24 hours to two days • More than two days

Example 9–B

Identifying Level of Measurement

Each level of measurement has the properties of any levels below it. Measure at the highest level of measurement appropriate for answering your question validly. The higher the level of measurement, the more specific the information gathered.

RATIO	Will measuring establish the proportion of differences between two or more objects that will be measured with this item? *If yes: The item meets critera for a ratio measure. Ratio is the highest level of measurement.* *If no: The item does not meet the criteria. Move down one level to see if this is an interval measure.*
INTERVAL	Will measuring establish the differences in equal units between two or more objects that will be measured with this item? *If yes: The item meets critera for a interval measure. Move up one level to see is it is a ratio measure.* *If no: The item does not meet the criteria. Move down one level to see if this is an ordinal measure.*
ORDINAL	Will measuring establish higher or lower (or better or worse), between two or more objects that will be measured with this item? *If yes: The item meets critera for a ordinal measure. Move up one level to see is it is a interval measure.* *If no: The item does not meet the criteria. Move down one level to see if this is a nominal measure.*
NOMINAL	Will measuring establish the differences between two or more objects that will be measured with this item? *If yes: The item meets critera for a nominal measure. Move up one level to see is it is a ordinal measure.* *If no: The item does not meet the criteria. The statement is not a measure.*

Example 9–C

Statements Written at Different Levels of Measurement	
LEVEL OF MEASUREMENT	STATEMENT USING LEVEL OF MEASUREMENT
Ratio	There are two females to every male.
	The ratio of females to males is 2 to 1 (2:1).
	The scale of the model to the prototype is 1 foot to 1 yard.
	The factor for converting from miles to kilometers is about six-tenths.
Interval	Twenty more females are present than males.
	The prototype is three feet longer than the mass-produced item will be.
	The current temperature is 10 degrees higher than this morning's temperature.
	The office is 10 miles farther from the house than the gym is.
Ordinal	More females are present than males.
	The prototype is longer than the mass-produced item will be.
	The current temperature is higher than this morning's temperature.
	The office is farther from the house than the gym is.
Nominal	There is a different number of females present than males.
	The prototype length is different from the length the mass-produced item will be.
	The temperature has changed since this morning.
	The distance from the house to the office from the house is different from the distance from the house to the gym.

Example 9–D

Data Collection Form

Survey of Child In Care

Date: ____/____/_____ (mm/dd/yyyy)   ID Number: _____ ____

1. When is your birthday? What year were you born? ___/___/_____(mm/dd/yyyy)
2. How long have you been staying where you live now? _____ Years _____Months

   Probe: How old were you when you came to live here (this time)?

   Were you living here on your last birthday? How about the birthday before that?

   Was it warm or cold outside? Do you remember what month it was?
3. How often does your worker come to meet with you? (About how often?)

   Check one box to indicate the answer than comes closest to your opinion.

   ◯ More often than once a month   ◯ Once a month   ◯ Less often than once a month
4. Would you say you worker is very helpful, somewhat helpful, not very helpful, or not at all helpful to you in getting adjusted to things where you live right now? (Check one box)

   ◯ Very helpful   ◯ Somewhat helpful   ◯ Not very helpful   ◯ Not at all helpful
5. What are the three best things about where you are living now?

6. What are some things that could be done to make things better for you where you are living now?

Note: Example 9-D shows examples of questionnaire data collection items. It does not show a complete data collection form.

Leaders

Developer-Manager and Reviewer

Skills

- Employing research methods and process.

- Designing and developing questionnaires.

Barriers

- Listing all questions that are of interest in the initial drafting of questionnaires.

- Trying to include all items of interest early in survey development, such as in a pilot questionnaire, can waste valuable time and cloud the survey's focus.

Binder Task

Insert into the binder the drafts of the forms, preserving information on changes and reasons for them.

Application of Tasks

Jean-Claude and Shaniqua scoured their files, journals, and the Internet, and they spoke to colleagues at length to determine what questions to include in the survey. They located a few forms with items that could be modified for the current project, but they concluded that they would have to develop most items. Both of the developer-managers were aware of the work required to develop new items. In addition to the time spent carefully crafting the items, this entails pretesting, revising, obtaining objective judgments, and revising again. They knew that serious mistakes could be made when previously constructed, validated data collection instruments and forms are incorporated in their entirety. Such items are valid only for the populations for which the forms have been used, and items borrowed for a new form may have been constructed for purposes entirely different from the current study's purposes. Jean-Claude and Shaniqua knew that objective colleagues would examine the items they adapted and compare the purposes and stated objectives of the prior and current studies, just as newly developed items would be judged.

Jean-Claude and Shaniqua also considered the crosscutting variables of site, supervisor, point in process, and workflow. As the director of the information technology division, Shaniqua had the codes used to identify site and supervisor. Such geographic and organizational codes are commonly part of electronic databases. They allow tracking of client assignments and workloads and can make it easier to reach clients and caseworkers in an emergency.

Measuring workflow and identifying key points in the work process were more complicated. Because a precise measurement of the work process would be a separate and major project on its own, Jean-Claude decided to use information he had already gathered about Stellar Agency's policies and procedures regarding work process. Jean-Claude's airm was to develop ideas about where barriers might arise, so that information could be gathered about how staff circumvented or solved problems in processes. He also considered two additional approaches: asking the individuals who do the work what they know about the workflow process, and observing the work while it is being performed and recording the process.

In addition to devising their surveys, the developer-managers used the information available to them to map key aspects of the service delivery process. They identified junctures at which workers open cases, delivered services, passed clients from one individual or division to another, and closed cases. Using a flowchart, they sketched the decision points and timelines associated with these tasks. The flowcharts were very useful in helping them understand the written policies and procedures. (Note: Many books and websites are available on how to construct a flowchart.)

After mapping the process, Shaniqua and Jean-Claude devised a scheme to code points and specific processes that form the segments between points. Next, they developed data collection forms to capture observations that measure the time (minutes, hours, or days) that it takes for a case to move from one point to another in the process. These observations might be through direct observation or time stamping of case materials on completion of tasks. They also developed forms to capture and code observations.

In addition, Jean-Claude and Shaniqua developed forms to collect information from workers on issues known to slow work processes, such as not getting needed resources. For example, malfunctioning copiers can impede having materials ready for court appearances, nonfunctioning voicemail can delay workers' knowing that clients are trying to reach them, and supervisors' not responding in a timely manner can leave workers without needed guidance.

Examples

Examples D through H are examples of the data collection forms that Stellar Agency developed for its study. Please note that these forms show examples of data collection items. They do not represent complete data collection forms.

Example 9–E

Format of a Case Review Data Collection Form

Stellar Agency Review of Child Record

Date: _____/_____/_____ (mm/dd/yyyy)   ID Number: _____

1. When did the child come into care (for this placement)? __/__/____ (mm/dd/yyyy)
2. When was the child's current worker assigned to the child's case? __/__/____ (mm/dd/yyyy)
3. How many placement moves has this child experienced? _____ Moves
4. Is the date of the most recent assessment or assessment update...
    ○ More than 12 months ago   ○ 12–6 months ago   ○ About 6 months ago   ○ Less than 6 months ago
5. Is this child receiving services according to the most recent service plan in the record?
    ○ Yes   ○ No   ○ No service plan   ○ Can't tell
6. Are the birthparents receiving services according to the most recent service plan in the record?
    ○ Yes   ○ No   ○ No service plan   ○ Can't tell
7. In your judgment, is this case moving toward the specified case outcome?
    ○ Yes ∫ Answer 7a   ○ No ∫ Skip to Question 8
7a. Is it moving in a reasonably timely manner, given the case particulars?
    ○ No ∫ Answer 7b   ○ Yes ∫ Skip to Question 8
7b. Please describe your judgment of the main reason for the slow pace of movement:

8. Is parental visitation occurring as specified in the most recent parental visitation plan?
    ○ Yes   ○ No   ○ No visitation plan   ○ Can't tell
9. Is sibling visitation occurring as specified in the most recent sibling visitation plan?
    ○ Yes   ○ No   ○ No visitation plan   ○ Can't tell   ○ Child doesn't have siblings

Example 9–F

Facility Observation Data Collection Form

Observation of Facility Site

Date: _____/_____/_____ (mm/dd/yyyy)   ID Number: _____

1. How many places for fire extinguisher holders are empty? (0 = All fire extinguishers are empty)

    # _____

2. Is at least one copier in working order at the site?   ○ Yes   ○ No

3. Does every worker who interviews clients in the office have a private space to conduct interviews?

    ○ Yes   ○ No

4. What is the highest number of rings that you recall hearing before the call was answered or the caller hung up?

    # _____

5. Is an accessible bathroom available to males and females on every level?   ○ Yes   ○ No

6. Are all bathrooms clean?   ○ Yes   ○ No

7. Is the waiting room well lighted?   ○ Yes   ○ No

8. Is the waiting room clean?   ○ Yes   ○ No

9. Are there any tripping hazards?   ○ Yes   ○ No

10. Are clerical work areas ergonomically safe?   ○ Yes   ○ No

Example 9–G

Respondent-Administered Questionnaire Data Collection Form

Stellar Agency 2000 Survey of Direct Service Workers

We want to learn how well resources reach those who need them. Please complete this questionnaire, put it in the envelope provided, and place it in any of the drop boxes placed throughout the organization. Thank you in advance for your participation.

1. How long have you worked for the Stellar Agency? (Please check only one box)

    ○ Less than 1 year   ○ 1–5 years   ○ 6–10 years   ○ 11–20 years   ○ More than 20 years

2. What types of services do you work in or provide in your current position? (Please check all that apply)

    ○ Child protective services   ○ Family preservation   ○ Family reunification
    ○ Foster care   ○ Adoption   ○ Independent living
    ○ Preparation for adult living   ○ Other (specify below)

4. How long have you been in your current position? (Please check only one box)

    ○ Less than 1 year   ○ 1–5 years   ○ 6–10 years   ○ 11–20 years   ○ More than 20 years

5. Thinking about last month only, for each of the resources listed below, indicate whether you usually received them or had them available: the time you first needed them, a few days after you first needed them, a week or two after you first needed them, longer than two weeks after you needed them, or whether you never received them at all. (Please check one box for each resource)

Type of Resource	At Time First Needed	A Few Days After First Needed	A Week or Two After First Needed	Longer Than Two Weeks After First Needed	Never Received at All	NA: Not Needed/ Requested
Office/clerical support	○	○	○	○	○	○
Computer applications	○	○	○	○	○	○
Supervisor input	○	○	○	○	○	○
Office supplies	○	○	○	○	○	○
Administrative input	○	○	○	○	○	○
Resource for client	○	○	○	○	○	○
Client information	○	○	○	○	○	○
Client cooperation	○	○	○	○	○	○

Example 9–G (continued)

5a. Are there are resources you have needed or requested that are not specified in Question 5?

　○ Yes　　　　　　　○ No (Skip to Question 6)

5b. Please describe the most important resource you have need or requested below:

5c. Please indicate whether you received or had available the resource described in Question 5b: at the time you first needed it, a few days after you first needed it, one or two weeks after you first needed it, longer than two weeks after you needed it, or never received it at all.

　○ At time first needed　　　　○ A few days after　　　　○ A week or two after
　○ Longer than two weeks after　○ Never received

6. Still thinking about the past month only, indicate the helpfulness to you of each resource after it was received. (Please check one box for each resource)

TYPE OF RESOURCE	VERY HELPFUL	SOMEWHAT HELPFUL	NOT VERY HELPFUL	NOT AT ALL HELPFUL	NA: NOT REQUESTED OR NOT RECEIVED
Office/clerical support	○	○	○	○	○
Computer applications	○	○	○	○	○
Supervisor input	○	○	○	○	○
Office supplies	○	○	○	○	○
Administrative input	○	○	○	○	○
Resource for client	○	○	○	○	○
Client information	○	○	○	○	○
Client cooperation	○	○	○	○	○
Other (specified in 5b)	○	○	○	○	○

7. Please rate the quality of the resources that you usually receive, using Excellent, Good, Fair, or Poor. (Please check one box for each resource)

TYPE OF RESOURCE	EXCELLENT	GOOD	FAIR	POOR	NA: NOT A RESOURCE USED
Office/clerical support	○	○	○	○	○
Computer applications	○	○	○	○	○
Supervisor input	○	○	○	○	○
Office supplies	○	○	○	○	○
Administrative input	○	○	○	○	○
Resource for client	○	○	○	○	○
Client information	○	○	○	○	○
Client cooperation	○	○	○	○	○
Other (specified in 5b)	○	○	○	○	○

8. Thinking again about the past month only, overall, how much difficulty were the issues presented by the cases assigned to you: more than expected, about as much as expected, or less than expected?

　○ More than expected　　　　○ About as much as expected　　　　○ Less than expected

9. Is there anything that someone in Stellar Agency could do to help direct service workers improve their timeliness in working with clients?

　○ Yes　　　　　　　○ No (Skip to Question 11)

10. Please describe what someone in Stellar Agency could do:

11. If you have any other comments, please write them here:

Thank You for Your Participation

Note: This example coordinates with Example 10-A.

Example 9–H

Interviewer-Administered Questionnaire Data Collection Form

Stellar Agency 2000 Survey of Direct Service Workers

1. How long have you worked for the Stellar Agency? (PLEASE CHECK ONLY ONE BOX)

   01   Less than 1 year
   02   1–5 years
   03   6–10 years
   04   11–20 years
   05   More than 20 years
   99   REF

2. I'm going to read a list of services that can be provided by Stellar Agency direct service workers. For each one, please tell me whether or not you provide that service in your current position.

   Do you provide (READ SERVICE)… (CIRCLE 01 IF RESPONDENT PROVIDES THE SERVICE, 00 IF NOT)

01	00	Child protective services
01	00	Foster care
01	00	Preparation for adult living
01	00	Family preservation
01	00	Adoption
01	00	Independent living
01	00	Family reunification
01	00	Other (RECORD VERBATIM IN 2a)

2a. RECORD R's DESCRIPTION OF "OTHER" VERBATIM (CODE UP TO 3)

3. How long have you been in your current position? (CIRCLE ONE)

   01   Less than 1 year
   02   1–5 years
   03   6–10 years
   04   11–20 years
   05   More than 20 years
   88   DK
   99   REF

4. Thinking about last month only, indicate whether you usually received or had available when needed each of the resources listed below: At the time you first needed them, a few days after you first needed them, a week or two after you first needed them, longer than two weeks after you needed them, or whether you never received them at all. (CIRCLE ONE NUMBER SET ON EACH LINE)

Type of Resource	At Time First Needed	A Few Days After First Needed	A Week or Two After First Needed	Longer Than Two Weeks After First Needed	Never Received at All	NA: Not Needed/ Requested	REF
Office/clerical support	01	02	03	04	05	88	99
Computer applications	01	02	03	04	05	88	99
Supervisor input	01	02	03	04	05	88	99
Office supplies	01	02	03	04	05	88	99
Administrative input	01	02	03	04	05	88	99
Resource for client	01	02	03	04	05	88	99
Client information	01	02	03	04	05	88	99
Client cooperation	01	02	03	04	05	88	99

4a. Are there resources you needed or requested last month that are not specified in Question 4?

   1   Yes
   0   No (SKIP TO Question 5)
   8   Don't remember
   9   REF

4b. Please describe the most important other resource you needed or requested last that is not listed in

Example 9–H (continued)

Question 4.

4c. Please indicate whether you received or had available the resource described in Question 5b: at the time you first needed it, a few days after you first needed it, one or two weeks after you first needed it, longer than two weeks after you needed it, or whether you never received it at all.

1   At time first needed

2   A few days after

3   One or two weeks after

4   Longer than two weeks after

5   Never receive

8   Don't remember

9   REF

5. Still thinking about the past month only, indicate the helpfulness to you of each resource after you received it: (PLEASE CIRCLE ONE NUMBER SET FOR EACH RESOURCE)

TYPE OF RESOURCE	VERY HELPFUL	SOMEWHAT HELPFUL	NOT VERY HELPFUL	NOT AT ALL HELPFUL	NA: NOT REQUESTED OR NOT RECEIVED	REF
Office/clerical support	04	03	02	01	88	99
Computer applications	04	03	02	01	88	99
Supervisor input	04	03	02	01	88	99
Office supplies	04	03	02	01	88	99
Administrative input	04	03	02	01	88	99
Resource for client	04	03	02	01	88	99
Client information	04	03	02	01	88	99
Client cooperation	04	03	02	01	88	99

6. Please rate the quality of the resources that you usually receive, using excellent, good, fair, or poor.

CIRCLE ONE NUMBER SET ON EACH LINE

TYPE OF RESOURCE	EXCELLENT	GOOD	FAIR	POOR	NA: NOT A RESOURCE USED	REF
Office/clerical support	04	03	02	01	88	99
Computer applications	04	03	02	01	88	99
Supervisor input	04	03	02	01	88	99
Office supplies	04	03	02	01	88	99
Administrative input	04	03	02	01	88	99
Resource for client	04	03	02	01	88	99
Client information	04	03	02	01	88	99
Client cooperation	04	03	02	01	88	99

7. Thinking again about last month only, overall, how difficult were the cases assigned to you: more than you expected, about as much as you expected, or less than you expected?

3   More than expected

2   About as much as expected

1   Less than expected

8   No opinion

9   REF

8. Is there anything that someone in the Stellar Agency could do to help direct service workers improve their timeliness in working with clients?

1   Yes

0   No (SKIP TO QUESTION 10)

8   DK

9   REF

Example 9–H (continued)

9. Please describe what someone in the Stellar Agency could do: (RECORD VERBATIM, CODE UP TO 5)

10. Do you have any comments? (RECORD VERBATIM, CODE UP TO 3)

Interviewer Comments/Notes: (CODE UP TO 3)

Interviewer Signature:_____ ID: _ _ _ _____ _

Date Interview Completed: ____/____/_____ mm/dd/yyyy

Note: This example coordinates with Example 13-B. Examples 9-I through 9-M illustrate data collection forms that Stellar Agency might use for its planned review.

Example 9–I

Data Collection: Types of Service or Equipment Available to Stellar Agency Staff

In this example, the variable, service or equipment, has been operationalized by breaking the variable into essential components (telephone, voicemail, and copier). Each item is measured separately. One data source for some of these measures is business records that indicate what equipment should be available. The researchers developed the following items, however, to obtain information from staff by a trained interviewer. The adequacy of each component is measured with different items in Example G.

This data collection form is designed to learn from direct service staff how available others are to them for work-related assistance. Administrated by a trained interviewer, the developer-managers developed it using the same principles that guided the development of Examples H and J. The items are precoded, that is, number codes are entered to represent responses, and data collectors are instructed how to code in all capital letters. When the respondent completes the form himself or herself, it is generally preferable to provide boxes for the respondent to check instead of using printed codes.

Note the items that ask for information other than the number of "wait" days. These items put the number of days in a context. This context will be constructed through data and statistical analyses, possibly from constructs from several variables. This issue is examined further in Step 12.

Interviewer-Administered Data Collection Items

Availability of Service/Equipment

I am going to read some equipment, services, and supports that are often found or needed in offices. As I read each, please say yes if you ever need the equipment, service, or support, or say no if you never need it.
(CIRCLE THE NUMBER)

A. Telephone

  1 Yes   2 No (GO TO B.)

1. Do you usually have access to an office telephone when you need one?

  1 Yes (GO TO B.)   2 No

2. How many workdays have you been without ready access to an office telephone when you needed one? _____ days

Example 9–I (continued)

3. Were the days without telephone access consecutive, or were there some days in between when you had access to an office telephone?

    1 Consecutive days   2 Days in between with access to a telephone

B. Voicemail

    1 Yes   2 No (GO TO C.)

1. Do you receive all of your telephone messages?

    1 Yes (GO TO C.)   2 No

2. Was there previously an adequate message system in place for you?

    1 Yes   2 No (GO TO C.)

3. How many weeks/days has it been since you had an adequate message system?

    (CONVERT TO DAYS; USE ZERO "0" FOR NONE) _____ days

4. How many weeks/days has it been since you had access to a telephone service that was adequate for you? (CONVERT TO DAYS; USE ZERO "0" FOR NONE) _____ days *

5. Overall, how adequate would you say the telephone service usually avaliable to you is? **

    4 Very adequate   3 Somewhat adequate   2 Not very adequate   1 Not at all adequate

C. Photocopier

    1 Yes   2 No

1. Do you have access to an office copier when you need one?

    1 Yes   2 No

2. Do you have access to a copy service?

    1 Yes   2 No

3. Did you previously have access to a copier or a copy service that was adequate for you?

    1 Yes   2 No

4. How many weeks/days has it been since you had access to a copier or a copy service that was adequate for you? (CONVERT TO DAYS; USE ZERO "0" FOR NONE) _____ days

*,** These questions are optional.

Example 9–J

Data Collection: Staff Perceptions of Support Available from Other Stellar Agency Staff and from Clients

Interviewer-Administered Data Collection Items

Perceptions of Support

I am going to ask about some of the people on whom you might count for input (advice, assistance, or some other type of help) to get your job done. As I identify each person, please say yes if you ever need advice, assistance, or some other type of help from that person. Say no if you never need this type of input from that person. (CIRCLE THE NUMBER)

A. Your direct supervisor

    1 Yes   2 No (GO TO B.)

1. About how many work days do you usually wait for your direct supervisor's needed input?

    _____ days   (USE ZERO "0" FOR NONE)

Example 9–J (continued)

2. How satisfied are you usually with the effort your direct supervisor makes to provide the input you need to do your job?

   4 Very satisfied   3 Somewhat satisfied   2 Not very satisfied   1 Not at all satisfied

3. What generally happens with regard to the input that you need from your direct supervisor? Do you receive that input?

   4 Always   3 Usually   2 Not very often   1 Never

B. Your supervisor's direct supervisor

   1 Yes   2 No (GO TO C.)

1. About how many work days do you usually wait for input from your direct supervisor's supervisor? _____days   (USE ZERO "0" FOR NONE)

2. How satisfied are you usually with effort your direct supervisor's supervisor makes to provide the input you need to do your job?

   4 Very satisfied   3 Somewhat satisfied   2 Not very satisfied   1 Not at all satisfied

C. Peers who do the same type of work that you do

   1 Yes   2 No (GO TO D.)

1. About how many workdays do you usually wait for input from peers who do the same type of work that you do? _____ days (USE ZERO "0" FOR NONE)

2. How satisfied are you usually with input from peers who do the same type of work that you do?

   4 Very satisfied   3 Somewhat satisfied   2 Not very satisfied   1 Not at all satisfied

3. What generally happens with regard to the input that you need from peers who do the same type of work that you do? Do you receive that input?

   4 Always   3 Usually   2 Not very often   1 Never

D. Colleagues and peers who do a different type of job:

   1 Yes   2 No (GO TO E.)

1. About how many workdays do you usually wait for input from colleagues and peers who do a different job? _____days   (USE ZERO "0" FOR NONE)

2. How satisfied are you usually with input from colleagues and peers who do a different job?

   4 Very satisfied   3 Somewhat satisfied   2 Not very satisfied   1 Not at all satisfied

3. What generally happens with regard to the input that you need from your colleagues and peers who do a different job? Do you receive that input?

   4 Always   3 Usually   2 Not very often   1 Never

E. Do you ever need your direct service clients to provide something so that you can proceed with their services (assessment, planning, or intervention)?

   1 Yes   2 No

1. About how many workdays do you usually wait for input from clients?

   _____days   (USE ZERO "0" FOR NONE)

2. Do you usually have to ask for input from direct service clients more than once?

   1 Yes   2 No

3. Usually, how useful is the input that they initially provide to you?

   4 Very usefull   3 Somewhat usefull   2 Not very usefull   1 Not at all usefull

4. Generally, how satisfied are you with the efforts that your direct service clients make to provide you with the input you requested?

   4 Very satisfied   3 Somewhat satisfied   2 Not very satisfied   1 Not at all satisfied

Example 9–K

Example 9-K integrates the questions presented in Examples 9H, and 9-J into a single interview form.

Complete Data Collection Form

Interview of all Stellar Agency Direct Service Workers

INTERVIEW DATE: ____/____/_____   ID NUMBER: 0 2 4 8

1. *I am going to read some equipment, services, and supports that are often found or needed in offices. As I read each, please say "yes" if you ever need the equipment, service, or support, or say "no" if you never need it. (CIRCLE THE NUMBER)*

A. *Telephone*

  1 Yes   2 No (GO TO B.)

  (1) Do you have access to an office telephone when you need one?
     1 Yes (GO TO B.)   2 No

  (2) How many workdays have you been without ready access to an office telephone?
     _____days

  (Alternative Questions: Has it been longer or shorter than five workdays? Has it been longer or shorter than two weeks? About how many of days in those weeks were days you were scheduled to work?)

B. *Voicemail*

  1 Yes   2 No (GO TO C.)

  (1) Do you receive all of your telephone messages?
     1 Yes (GO TO C.)   2 No

  (2) Was there an adequate message system previously in place for you?
     1 Yes   2 No (GO TO C.)

  (3) How many weeks or days has it been since you had an adequate message system?
     (CONVERT TO DAYS; USE ZERO "0" FOR NONE) _____days

C. *Copier*

  1 Yes   2 No (GO TO 2.)

  (1) Do you have access to an office copier when you need one?
     1 Yes (GO TO D.)   2 No

  (2) Do you have access to a copy service?
     1 Yes (GO TO D.)   2 No

  (3) Did you previously have access to an adequate copier or a copy service?
     1 Yes   2 No (GO TO 2.)

  (4) How many weeks or days has it been since you had access to an adequate copier or a copy service?
     (CONVERT TO DAYS; USE ZERO "0" FOR NONE) _____days

2. *Now, I am going to ask about some of the people you might count on for input (advice, assistance, or some other type of help) to get your job done. As I identify each person, please say yes if you ever need advice, assistance, or some other type of help from that person. Say no if you never need this type of input from that person.*

A. *Your direct supervisor:*

  1 Yes   2 No (GO TO B.)

  (1) About how many workdays do you usually wait for your direct supervisor's needed input?
     _____ days (USE ZERO "0" FOR NONE)

  (2) How satisfied are you usually with the effort your direct supervisor makes to provide the input you need to do your job?

     4 Very satisfied   3 Somewhat satisfied   2 Not very satisfied   1 Not at all satisfied

  Follow-up Question: Is your response closer to "somewhat satisfied" or "not very satisfied"?

  (3) What generally happens with regard to the input that you need from your direct supervisor?

Example 9–K (continued)

Do you receive that input?

| 4 Always | 3 Usually | 2 Not very often | 1 Never |

Follow-up Question: Is your response closer to usually or not very often?

B. *Your supervisor's direct supervisor*

1 Yes   2 No (GO TO C.)

(1) About how many workdays do you usually wait for input from your direct supervisor's supervisor? _____ days (USE ZERO "0" FOR NONE)

(2) How satisfied are you usually with the effort your direct supervisor's supervisor makes to provide the input you need to do your job?

| 4 Very satisfied | 3 Somewhat satisfied | 2 Not very satisfied | 1 Not at all satisfied |

Follow-up Question: Is your response closer to "somewhat satisfied" or "not very satisfied"?

(3) What generally happens with regard to the input that you need from your direct supervisor's supervisor? Do you receive that input?

| 4 Always | 3 Usually | 2 Not very often | 1 Never |

Follow-up Question: Is your response closer to "usually" or "not very often"?

C. *Peers who do the same type of work that you do*

1 Yes   2 No (GO TO D.)

(1) About how many workdays do you usually wait for input from peers who do the same type of work that you do? _____ days (USE ZERO "0" FOR NONE)

(2) How satisfied are you usually with input from peers who do the same type of work that you do?

| 4 Very satisfied | 3 Somewhat satisfied | 2 Not very satisfied | 1 Not at all satisfied |

Follow-up Question: Is your response closer to "somewhat satisfied" or "not very satisfied"?

(3) What generally happens with regard to the input that you need from peers who do the same type of work that you do? Do you receive that input?

| 4 Always | 3 Usually | 2 Not very often | 1 Never |

Follow-up Question: Is your response closer to "usually" or "not very often"?

D. *Colleagues and peers who do a different type of job:*

1 Yes   2 No (GO TO E.)

(1) About how many workdays do you usually wait for input from colleagues and peers who do a different job? _____ days (USE ZERO "0" FOR NONE)

(2) How satisfied are you usually with input from colleagues and peers who do a different job?

| 4 Very satisfied | 3 Somewhat satisfied | 2 Not very satisfied | 1 Not at all satisfied |

Follow-up Question: Is your response closer to "somewhat satisfied" or "not very satisfied"?

(3) What generally happens with regard to the input that you need from colleagues and peers who do a different job? Do you receive that input?

| 4 Always | 3 Usually | 2 Not very often | 1 Never |

Follow-up Question: Is your response closer to "usually" or "not very often"?

E. *Do you ever need your direct service clients to provide something so that you proceed with their services (assessment, planning, or intervention)?*

1 Yes   2 No

Example 9–K (continued)

(1) About how many workdays do you usually wait for input you need from clients?
_____ days (USE ZERO "0" FOR NONE)

(2) Do you usually have to ask for input from direct service clients more than once?
   1 Yes   2 No

(3) Generally, how useful is the input that they initially provide to you?

   4 Very useful   3 Somewhat useful   2 Not very useful   1 Not at all useful

Follow-up Question: Is your response closer to "somewhat useful" or "not very useful"?

(4) Generally, how satisfied are you with the efforts your direct clients make to provide the input you have requested?

   4 Very satisfied   3 Somewhat satisfied   2 Not very satisfied   1 Not at all satisfied

Follow-up Question: Is your response closer to "somewhat satisfied" or "not very satisfied"?

3. Do you have any additional comments?

Interviewer Comments/Notes

Interviewer Name (print):_____

Interviewer Signature:_____

Additional Reading and Reference

Sudman, S., & Bradburn, N. (1990). *Asking questions: A practical guide to questionnaire design.* San Francisco: Jossey-Bass.

Step 10:

Develop Training Materials and Train Interviewers

Task

- Begin to develop materials to train the data collectors. The training materials will be used later, but it is important to begin the development of training materials now so that they will be available for Step 14.

Definition

QXQ *specs*, or *question-by-question specifications*, are the directions for obtaining data on each item on a data collection form. QXQ specs also explain the purpose of each item on the form. QXQ specs may provide additional nonbiasing probes, describe how to identify a sample member, or explain how to code completed collections, refusals, or invalid sample selections. Example 10-A provides QXQ specs for the questionnaire in Example 9-F.

More Information

The precision of data collection determines the accuracy of the measurements and affects the validity of the results. The skillfulness of the data collectors determines how accurately they will collect data, and the training and briefings they receive will affect their accuracy.

Measurement begins with defining the phenomena, creating appropriate items, and putting the items into a form that conforms the plan for capturing the measures. Those tasks were covered in Steps 2 and 9. No matter how carefully these tasks were performed, however, the accuracy of the measures (and hence, the validity of the findings) depends on the accuracy of the captured data. If the data collection process includes bias, inaccuracies, or carelessness, painstaking care in designing and developing measures will not result in valid findings.

There are two types of training materials: (1) general tips on unbiased, valid data collection, and (2) training on the use of the data collection forms specific to a project. General data collection training and guidance topics include:

- Common data collection errors:

  - using the wrong sample member or believing that it is not important which sample member is included;

  - deviating from the wording when interviewing a respondent;

  - using words, gestures, or facial expressions that lead a respondent to answer a question in a specific way;

  - expressing opinions, before or during data collection, that hint at the interviewer's opinions in any area;

  - using leading probes, which could hint at the socially desirable response for a question;

  - using ineffective probes or not probing enough—that is, not obtaining responses that enable the data collector to form a mental picture of the response to an open-ended question;

– not recording a response to an open-ended item verbatim, so that the mental picture of the phenomenon described by the respondent will not be recorded for the next measurement tasks;

– transposing numbers or letters; and

– losing one's place in a series of data while copying them.

- Ways to avoid common data collection errors:

   – training on data collection tools that includes role playing for data collectors, in which they use the tools in simulated situations; and

   – taking care to familiarize interviewers with the points at which errors are likely to occur, so they can perform those tasks with heightened awareness and great care. Training is essential, because every step in measuring intangible phenomena, and even tangible phenomena, introduces the possibility of error.

Example 10-B outlines the knowledge and abilities required for various types of data collection methods. This information should guide the of the data collectors who will be selected and assigned to projects.

Leaders

Developer-Manager and Reviewer

Skills

- Employing research methods and process.

- Designing and developing questionnaires.

Barrier

- Becoming inattentive, and as a result, making errors. This step involves a high level of detailed, though not difficult, work. Alertness is essential.

Binder Task

Draft the QXQ specs for the data collection forms. If the questionnaire is changed, this document also must be changed.

Application of Tasks

Ruby Brown, social work training specialist, is an experienced interviewer and data collector who has worked on many research projects, both prior to joining Stellar Agency's staff and since joining the agency. She has experience collecting data through interviews, field observations, and the extraction of information from written materials. Ruby is responsible for developing the training materials, training the data collectors, and supervising the data collection. (A description of adult learning and training principles in beyond the scope of this guidebook. Please consult the suggested reading list at the end of Step 10 for guidance.)

Examples

Example 10–A

Question-by-Question (QXQ) Specifications (Interviewer-Administered Questionnaire)	
2000 Survey of Direct Service Workers QXQ Specifications	
General Instructions	*Write (X) to indicate each probe.*
Site	*Write the site letter of R's work location in the top, right-hand corner of each form page.*
Form Number	*Assign a number to R and form. Write the number on the form below the site letter.*
1. How long have you worked for the Stellar Agency?	*Circle 01 for yes or 00 for no.*
2. What types of services do you work in or provide in your current position?	*Read each service. Circle 01 if R indicates he or she provides it; circle 00 if R does not provide the service.*
2a. OTHER	*Record verbatim.* *Probe: Tell me more about that. Probe until you can visualize R performing the service.*
3. How long have you been in your current position?	*Reread the item and categories to probe.*

4. Thinking about this past month only, indicate whether you usually received or had available when needed each of the resources I am going to read: at the time you first needed them, a few days after you first needed them, a week or two after you first needed them, longer than two weeks after you needed them, or whether you never received them at all.	*Use the past whole calendar month. Example: If it is any time in January, use December; if any time in February, use January; and so on.*
	 Read the complete item. Then, read the resources, followed by the response categories:
	     01 At the time you first needed them
	     02 A few days after you first needed them
	     03 A week or two after you first needed them
	     04 Longer than two weeks after you needed them
	     05 Never received them at all

4-1. Office/clerical support 4-2. Computer applications 4-3. Supervisor input 4-4. Office supplies 4-5. Administrative input 4-6. Resource for client 4-7. Client information 4-8. Client cooperation	*Read the resource. Then read the response categories.*
4a. Are there resources you have needed or requested that are not specified in Question 5?	*OTHER*
4b. Please describe the most important other resource you have need or requested below that is not listed in Question 4.	*Record verbatim.*
4c. Please indicate whether you received or had available the resource described in Question 5b: at the time you first needed it, a few days after you first needed it, one or two weeks after you first needed it, longer than two weeks after you needed it, or whether you never received it at all	*Reread the item and categories to probe.*
5. Still thinking about the past month only, indicate the helpfulness to you of each resource after you received it.	*04 Very helpful* *03 Somewhat helpful* *02 Not very helpful* *01 Not at all helpful* *Code "NA" if R did not request this resource during the past month*

Example 10–A (continued)

5-1. Office/clerical support	*Read the resource. Then read the response categories.*
5-2. Computer applications	
5-3. Supervisor input	
5-4. Office supplies	
5-5. Administrative input	
5-6. Resource for client	
5-7. Client information	
5-8. Client cooperation	
6. Please rate the quality of the resources that you usually receive, using excellent, good, fair, or poor.	*04 Excellent*
	03 Good
	02 Fair
	01 Poor
	Code "NA" if R has no past experience with this resource or cannot recall sufficiently to rate
6-1. Office/clerical support	*Read the resource. Then read the response categories.*
6-2. Computer applications	
6-3. Supervisor input	
6-4. Office supplies	
6-5. Administrative input	
6-6. Resource for client	
6-7. Client information	
6-8. Client cooperation	
7. Thinking again about the past month only, overall, how much difficulty were the issues presented by the cases assigned to you: more than expected, about as much as expected, or less than expected?	*"Past month" is the most recent calendar month with a name.*
	3 More than expected
	2 About as much as expected
	1 Less than expected
8. Is there anything that someone in the Stellar Agency could do to help direct service workers improve their timeliness in working with clients?	*Circle yes or no*
9. Please describe what someone in the Stellar Agency could do.	*Record verbatim. Probe until you can visualize the action and understand its purpose. Code up to five after the interview.*
10. Do you have any comments?	*Record verbatim. Code up to three after the interview.*
Interviewer Comments/Notes	*Record verbatim. Code up to three after the interview.*
Interviewer Signature and ID Number	*Write your ID number clearly on the lines provided. If your ID has fewer than four digits, enter leading zeroes.*
Date Interview Completed	*Enter the month and day in two digits. Enter the year in four digits.*

Note: R = respondent. This example coordinates with Examples 9-G and 13-A.

Example 10–B

Data Collectors' Required Knowledge and Abilities for Specific Data Collection Methods	
Data Collection Method	**Required Knowledge/Ability**
Case reading, collecting data from paper files	Know the normal order of cases or files
	Know the regular documentation methods
	Be accurate about detailed work
	Able to record numbers without transposing them
	Able to accurately perform repetitious tasks
	Able to code information in high agreement with other team members
Observation of events	Able to write without missing the phenomena targeted for measurement
	Able to code observations in high agreement with the other team members
Interviewer-administered in-person interviews	Know conventions of item reading, probing, controlling facial expressions and other nonverbal cues, completing questionnaires
	Able to use silence tactically
	Able to go through the specific questionnaire without long periods of nontactical silence
Telephone interviews	Know conventions of item reading, probing, and completing questionnaires, including knowing how to use computer-assisted applications
	Able to read quickly and distinctly
	Able to avoid periods of nontactical silence
Mail with respondent-administered forms	Able to answer questions frequently asked by respondents, to use tact when a question cannot be answered and the caller must be connected with someone else, and to ensure that the caller has been connected with a person who can answer the question before getting off the line
In-person interviews with respondent-administered forms	Able to be tactfully observant to discourage respondents conferring with one another or to prevent spoiled questionnaires from being added to those completed properly

Additional Readings and References

Deporter, B., & McPhee, D. (1996). *Quantum learning*. New York: DTP Trade Paperbacks.

Kolb, D. A. (1984). *Experimental learning: Experience as the source of learning and development*. Upper Saddle River, NJ; Prentice Hall PTR.

Kolb, D. A. (1999). *Learning style inventory*. Boston: TRG Hay/McBer.

McPhee, D. (1996). *Limitless learning*. Tuscon, AZ: Zephyr Press

STEP 11:

Design the Sample

Tasks

- Confirm that the population, the unit of analysis, time or geographical restrictions, and the types of cases to be specifically excluded from the population have been correctly defined and noted, referring to Step 5. Build on previous work to more specifically define the population and the population elements and units, any applicable time limitations, and other characteristics that disqualify units from selection, such as all cases open for more than a specific time period or all cases assigned to a specific worker, direct supervisor, geographic region, or agency.

- Consider carefully which characteristics of the population are important for the project. The important population parameter is the standard error, a measure of how much population members and elements are similar or dissimilar. The more dissimilar population members and elements are, the larger the sample size required. When assessing the homogeneity of a population, use a characteristic that is important to the results. As an example, location (rural or urban) is frequently an important characteristic in child welfare service delivery. As another example, the degree of the assigned worker (MSW, MA, BA, other master's, or other bachelor's) is considered important in many studies. Almost any characteristic, however, can be important in assessing homogeneity or heterogeneity.

- If measures for the population are not available, develop estimates. Rely on your familiarity with the data and the information generated from the data. Ask staff members who collect, manage, and store data for ideas. Ask personnel who use the information derived from the data, and factor their responses into the analysis. Ask individuals who provide services about any similarities or differences in the populations they serve.

- Compute or estimate standard error measures for all the sample statistics you will be using or estimating with new data collection.

- Decide which type of sample would work best. Some of the most frequently used sample types are discussed in the More Information section.

- Compute the sample size required for the boundaries on the error that you decide to impose. Seek guidance from a sampling expert about sample design, including the factors to consider when computing sample sizes.

- Determine the bound on error. This number is important when computing the required sample size to ensure the analysis will have sufficient power to support the conclusions and answer the questions.

Definitions

Bound on error of estimation is the percentage of sampling error the researcher is willing to tolerate. In research, 5% (for a 95% confidence level) is common. For some studies, the error rate may be larger or smaller (Scheaffer, Mendenhall, & Ott, 1979).

Heterogeneity refers to the level of dissimilarity in the elements of the population. All else being equal, heterogeneous populations require the larger sample selections than homogeneous populations.

Homogeneity refers to the level of similarity in the elements of the population. All else being equal, homogeneous populations allow the selection of smaller samples than heterogeneous populations.

Parameters are measures of a population's characteristics and correspond to a sample's statistics. *Parameter* and *statistic* mean the same thing, except that *parameter* refers to the population and *statistic* refers to the sample.

Standard error is the estimated measure of how similar or different the population members are with respect to important traits that will guide interpretation of findings. For example, when a political survey is designed for the purpose of producing findings applicable to the entire United States, the researcher will assumed that political party affiliation will reflect all the recognized parties in the country, and the study will have a relatively large standard error. In a survey of Chicago voters regarding an upcoming national election, for example, a researcher, knowing that Chicago voters typically give roughly equal votes to Democrats and Republicans in national elections, would expect a large standard error. If the researcher were designing a survey of Chicago voters prior to a local election, however, and she knew that in previous city and county elections, Democratic candidates routinely received far more votes than Republican candidates, she would expect a very small standard error.

Leaders

Developer-Manager, Data Analyst, and Reviewer. These leaders should consult with a sampling expert before finishing this step.

Skills

- Employing research methods and process.

- Designing and selecting samples.

More Information

Four sample design techniques allow for generalization to the population:

- Simple random sampling (SRS) is used when all population elements have an equal chance of being selected. The study does not specify any special characteristics that are of interest. SRS is feasible when all population elements are together in a format such as a database, and the elements have unique identifiers.

- Probability proportionate to size (PPS) is used when the researcher wants divisions within the sample to be in proportion to their proportions in the population. As with SRS, this design is feasible if all population elements are together in a database and have unique identifiers.

- Stratified random sampling is used to ensure that the sample reflects the characteristics of particular interest. For example, the sample may include cases closed in less than 12 months, after more than 12 months but less than 15 months, and after 15 or more months. The researcher would divide cases into three groups, and three separate SRS sample selections would be made, one for each stratum.

- Systematic sampling is used when the population elements appear on a list numbered 1 through *n* or when the sample can be selected from a receptacle (such as a file drawer) that presents or reveals all elements equally. When sampling from a file drawer of records, however, it is important to note that some records may be dog-eared or otherwise hidden and could be left out of the sample selection. In systematic sampling, the population size is divided by the required sample size. The result minus 1 equals the number of records to be skipped between selected sample members after a random start. If 1 is not subtracted from the result, the applied interval will not result in a selection of a sufficient number of sample members. This technique is best with a large population.

Three techniques of sampling do not allow generalizing to populations:

- availability or convenience sampling, which involves the selection of whomever or whatever is at hand;

- snowball sampling, which involves locating individuals with special characteristics of interest and asking them for referrals to other individuals with the same or similar traits; and

- quota sampling, in which the researcher identifies important strata and uses stratum guidelines to identify elements eligible for selection. Unlike PPS, however, the researcher selects sample members according to availability or convenience guidelines.

With regard to putting a bound on random sampling error, samples can be designed so that the probability of random sampling error is restricted. First, the evaluator must decide on a bound for error, frequently 5%. A higher bound can be tolerated in some instances, but in some cases, estimates of averages, quantities, or proportions must be very precise.

The actual error during data analysis (observed after the data have been collected) can be and often is smaller than the bound set during sample design. The error also can be larger than planned, particularly if the completion rate is low. If you anticipate that there may be a problem obtaining the number of expected completions, modify the design to select a larger sample. Setting the bound only keeps probable error in check.

Barriers

- A strong urge to skip this step if numbers or estimates for populations are not readily available. Doing so, however, is a mistake. Logically developed estimates are essential to calculations for corrections that may be necessary to generalize sample results.

- Difficulty in selecting a sample with any chance of being generalized to a population. This problem may arise when information is not in a database or on a numbered list.

Binder Task

Create a table to display the relevant population parameters. Leave a blank section in the table for the corresponding sample statistics, which will be created in Step 15.

Application of Tasks

When a project is complex or the agency staff does not have a background in sampling and statistics, the advice of a sample expert should be sought. Jean-Claude and Shaniqua, however, implemented a simple approach to making sampling decisions. Table 9 charts each step in their decisionmaking process.

TABLE 9: Stellar Agency: Steps Taken in the Sample Design	
Action	Steps/Resources
Consider which characteristics of the population are important.	Previous summaries of Stellar Agency's demographics show that the most important population characteristics vary across the four sites. Generally, Site A has the greatest variation (that is, it is the most heterogeneous), and Site B has the least variation. Sites C and D fall between these two sites. Appendix D provides the population characteristics.
Select the sample type.	Stellar Agency will use stratified random sampling. It will sample clients by site or region and type of client: individual, family, and group, with individuals further stratified by age, child (<18) and adult (>18). The agency will use stratified random sampling of direct service workers by site or region. (Although individual direct service workers would like feedback specific to their caseloads, it will not be practical to provide such information in the current project). The agency will use stratified random sampling of clerical workers will be by site or region, and for the organization, the study will have no stratification or sampling. The universe will be selected.
Decide on the bound on error.	5%
Compute sample sizes.	Stellar Agency computed sample sizes using the formulas shown in Step 11, Examples A and B (see Shaeffer et al. [1979] for other formulas).
Increase computed sizes to allow for ineligible selections and nonresponse.	Stellar Agency increased computed sample size by the conservative level of 30%, but may increase the sample size further, if indicated.

Examples

Example 11–A

Formula that Can Be Used to Estimate Sample Size

Estimating Sample Size (to estimate means, totals, proportions) using SRS, Population Size is Known

- $n = 357.71$
- sample size = 358, where,
  - B = bound on error of estimation = .05, or 5%;
  - N = population size = 3,375 = number of Stellar Agency individual clients, all ages;
  - n = estimated minimum sample size required, rounded to 358; and
  - p = variance, conservative for proportion in this example = .5. If estimating the mean, compute the variance of the population from the records or estimate from previous surveys
  - $q = 1\text{-}p$, i.e., the difference between 1 and the value of p.

After calculating their original sample size, Shaniqua and Jean-Claude remembered something they had read in the literature about the high nonreponse rate for surveys. They decided to hold the variance and error rates constant for their population, while increasing their sample by 30%, to $n = 465.032$, or 466 (see Shaeffer et al. [1979], p. 49).

$$n = \frac{Npq}{(N\text{-}1)\dfrac{B^2}{4} + pq}$$

$$n = \frac{(3375)\,(.25)}{(3374)\,(.000625) + .25}$$

$$n = 357.71, \simeq 358$$

Example 11–B

Stratified SRS: Computing Stratum Proportions for Allocating Total Sample Size to Strata

Example 11-A shows how to compute the total sample size. The size derived does not allow for analysis within strata. Below, the example uses the equation from Example 11-A to compute minimum sample sizes for each stratum. The formula applied to each stratum will sum to a number larger than that computed in Example 11-A. A separate formula for computing total and stratum sample sizes exists (see Schaeffer et al. [1979], p. 49).

1. Increase the sample size by 30% to allow for nonresponse: $n_1 \simeq 312$).

$$n_1 = \frac{600 \,(.25)}{(599)(.000625)+25} = 240.24$$

2. Increase the sample size by 30% to allow for nonresponse: $n_1 \simeq 380$).

$$n_2 = \frac{1080 \,(.25)}{(1079)(.000625)+.25} = 292.09$$

3. Increase the sample size by 30% to allow for nonresponse: $n_1 \simeq 360$).

$$n_3 = \frac{900 \,(.25)}{(899)(.000625)+.25} = 277.14$$

4. Increase the sample size by 30% to allow for non-response: $n_1 \simeq 346$).

As shown, the total number in a sample of individuals, stratifying by site or region, is 1,398. The stratum sizes are from Stellar Agency's demographics, as shown in Appendix D.

$$n_4 = \frac{795 \,(.25)}{(794)(.000625)+.25} = 266.33$$

Additional Reading and Reference

Scheaffer, R., Mendenhall, W., & Ott, L. (1979). *Elementary survey sampling* (2nd ed., Chps. 1–5, 8). North Scituate, MA: Duxbury Press.

Step 12:

Draft the Preliminary Data Analysis Plan

Tasks

- Review your notes from the previous steps, especially notes on the questions you need to answer (Steps 1 and 8), the variables you defined for answering the questions (Steps 2 and 5), the outcomes that will indicate whether outcomes are being achieved (Step 4), and the data collection forms (Step 9).

- Identify the likely cross-cutting variables. Write the first question at the top of a page. Under the question, describe which variables to use, including cross-cutting variables, and the procedures with which to analyze the variables, such as cross-tabulations and chi-square tests. Repeat this process for each question.

- Draft the preliminary data collection plan without technical jargon. After you have expressed precisely what the study needs to do, consult a manual for the statistical package that can convert the plain words of the plan into appropriate computer terminology.

More Information

The data analysis plan at this step is preliminary. Its purpose is to help the developer-managers ensure that they will have the data necessary to answer the questions and meet the project goals. Developing a preliminary data analysis plan helps verify that the study includes all the necessary variables and measures the variables at a level that will create the needed data. It allows the developer-manager to evaluate progress. At this point, it is possible to return to earlier steps and fill in gaps, but shortly, it will not be possible to do so.

Leaders

Developer-Manager, Data Analyst, and Reviewer

Skill

- Employing research methods and process.

Barriers

- Evaluation or research questions that were not well specified in Step 1.

- Use of measures that do not collect data for answering the questions, that is, poorly developed or missing items.

- Lack of clear connections between review questions, data collection items, and variables.

Binder Task

Insert the pages describing the variables and procedures for analysis from the Step 12 task into the project binder.

Look over the review, evaluation, and research questions. Which variables will provide the answers? Will you need to look at some of the variables by categories or values of the variable? If so, should you cross-tabulate or regress? Note the statistical actions you will take in the binder.

TABLE 10: Data Analysis Plan for Stellar Agency Quality Assurance Survey	
Data Item	**Narrative Instructions**
	"I am going to read some equipment, services, and supports that are often found or needed in offices. As I read each, please say yes if you ever need the equipment, service, or support, or say no if you never need the item."
Telephone 1 Yes 2 No	Create a duplicate of this variable.
	For the new variable, recode 2 to 0; calculate frequency distribution and statistics: number valid responses (n), mean (y), standard deviation (SD), and median.
Do you have access to an office telephone when you need one? 1 Yes 2 No	Create a duplicate of this variable.
	For the new variable, recode 2 to 0; code those who said no as 8 and declare that value missing.
	Calculate frequency distribution and statistics: n, y, SD, median; n should equal total number minus the cases recoded as 8.
How many workdays have you been without ready access to an office telephone? _____ days	Create a duplicate.
	Recode 2 to 0; code those who said no as 8 and declare missing.
	Calculate frequency distribution and statistics: n, y, SD, median; n should equal total number minus the cases recoded as 8.
	Get analysis of variance statistics.
	F test the three new variables by cross-cutting variables: site or region, point in work flow, and supervisor.

Application of Tasks

Joe Jones works in both the information technology division and the knowledge management division. He also is the data analyst for quality assurance projects. Joe, Shaniqua, and Jean-Claude drafted the preliminary data analysis plan together, after they reviewed Steps 1, 2, 4, and 9 to refresh their memories on the questions developed to guide the study, the variables defined from the questions, and the items developed to collect information for answering the questions. Table 10 illustrates the data analysis.

Based on the structure of their quality assurance study so far, the data analysis plan addressed the mean wait time for adequate access to a telephone. The same process could also be used for access to voicemail or a copier, with the narrative tailored to each, if appropriate to the study.

Practice Question

12-1. Study the beginning draft plan for data analysis, shown in Table 10. What can Joe, Jean-Claude, and Shaniqua do to determine which resource (telephone, voicemail, or copier) direct service workers perceive as having the worst performance?

Design and Document Database Layouts

Tasks

- Consult a professional with skills in creating, writing, and matching computer databases and data files. Data files include files based on previously collected and stored data and files that will be created.

- Get a copy of the file layout for the file.

- Create the files you need (see More Information).

- If more than one form is part of the review, the identifiers (also called the sorts/keys) for sample members must have the same name, data type, and width in all data files to link the files later.

Definitions

A *data dictionary* is a listing that defines the layout of a file to be created.

A *file layout* is a map of the fields that will contain the variables. The map specifies the following for each variable:

- Variable name

- Type—alphabetic or numeric

- Width—the number of spaces or placeholders it uses. For example, "12" uses two spaces, and "1.5" uses three spaces. This map section also specifies whether there are decimal spaces for numeric data and, if so, the number of such spaces.

Sort/key means the field or variable that will be used to sort the file and match file records to records in another file.

More Information

Often, an evaluation will use data collected from previously developed databases or data files. This is called *secondary analysis*. Case review information is often linked with administrative or other data in management information systems. For example, a study may use a child's demographics, family information, history of moves, data on length of stay, and worker assignment information. Files based on previously collected data are the frame for selecting the sample and provide the foundation for tracking the status of sample members. They are the basis for managing the sample and tracking the status of sample members throughout the project, and they establish the foundation for creating statistical program.

SPSS® (Statistical Package for the Social Sciences) is popular and often used in human service environments. SAS® (Statistical Analysis System) is also used often. Because personal computers are being used more extensively, many more applications can be used for the purposes described in this guidebook. Relational databases and spreadsheets now offer analysis and graphing functions that meet the needs of quality professionals. It is essential that the guides included with any application be followed closely.

Examples of useful files to create include:

- population records with sample records indicated;

- sample records with fields added for managing returns and completion status, and from which reports of completion rates can be generated; and

- files for entering data from returned forms, with sample record identifiers added so that analyses and results can be linked with sample member characteristics. This type of file allows the data analyst to report on any systematic differences between responding and nonresponding sample members. Based on this information, the data analyst can determine the extent to which results can be applied to the population and whether the data can be weighted for analysis, that is, whether proportions found in subsample members represent their presence in the population.

Leaders

Data Analyst, Developer-Manager, and Reviewer

Skills

- Employing research methods and process.

- Designing and selecting samples.

- Employing data analysis process.

 ## Barriers

- Mistakes in calculating spaces for field widths. Plan carefully. If you are not certain about exact widths, err on the side of too many spaces, rather than too few.

- Numbers stored as text. Make certain that the keys in different files are all of the same type. Otherwise, files cannot be matched and linked.

 ## Binder Task

Insert the file layout into the binder. Example B illustrates a file layout.

Application of Tasks

Joe, Shaniqua, and Jean-Claude planned the layouts for the databases. They determined

- how many variables the database would include;

- the order of the variables;

- the number assigned to each variable;

- the title of each variable, designed to make it easy to identify;

- the type of field representing each data element, such as numeric, alphanumeric, and so forth;

- the number of spaces, or field width, that each element's value would occupy when typed into the database;

- the column position of the beginning and ending space occupied by each variable; and

- any relevant restrictions or comments, such as how to code a missing value or what case to use when entering words.

Joe carried out the plan and created the layouts.

Examples

Example 13–A

This example illustrates operationalization and the process to construct items for data collection forms.

		Data Definition for Creating a Data File Layout				
		Stellar Agency 2000 Survey of Direct Service Workers				
		Data Definition				
POSITION	QUESTION #	VARIABLE/FIELD NAME	VARIABLE/FIELD TYPE	FIELD WIDTH	COLUMN BEGIN-END	RESTRICTIONS/COMMENTS
1	—	Site	Numeric	1	1–2	
2	—	Form	Numeric	3	3–5	001225
3	1	Lngwrk	Numeric	2	6–7	
4	2-1	Prvcps	Numeric	2	8–9	
5	2-2	Prvfcs	Numeric	2	10–11	
6	2-3	Prvpal	Numeric	2	12–13	
7	2-4	Prvfps	Numeric	2	14–15	
8	2-5	Prvadp	Numeric	2	16–17	
9	2-6	Prvils	Numeric	2	18–19	
10	2-7	Prvfrs	Numeric	2	20–21	
11	2-8	Prvoth	Numeric	2	22–23	
12	2a	Verb2a	Alpha	30	24–53	First 30 letters verbatim
13	2a-1	Oth2a1	Numeric	3	54–56	
14	2a-2	Oth2a2	Numeric	3	57–59	
15	2a-3	Oth2a3	Numeric	3	60–62	
16	3	Lngpos	Numeric	2	63–64	
17	4-1	Rcscler	Numeric	2	65–66	
18	4-2	Rcscomp	Numeric	2	67–68	
19	4-3	Rcssupv	Numeric	2	69–70	
20	4-4	Rcssupp	Numeric	2	71–72	
21	4-5	Rcsadms	Numeric	2	73–74	
22	4-6	Rcsclnt	Numeric	2	75–76	
23	4-7	Rcsinfo	Numeric	2	77–78	
24	4-8	Rcscoop	Numeric	2	79–80	
25	4a	Resoth4a	Numeric	1	81–81	
26	4b	Verb4b	Alpha	30	82–111	First 30 letters verbatim
27	4b	Oth4b	Numeric	2	112–113	
28	4c	Recrcs	Numeric	1	114–114	
29	5-1	Hlpcler	Numeric	2	115–116	
30	5-2	Hlpcomp	Numeric	2	117–118	
31	5-3	Hlpsupv	Numeric	2	119–120	
32	5-4	Hlpsupp	Numeric	2	121–122	
33	5-5	Hlpadms	Numeric	2	123–124	
34	5-6	Hlpclnt	Numeric	2	125–126	
35	5-7	Hlpinfo	Numeric	2	127–128	
36	5-8	Hlpcoop	Numeric	2	129–130	
37	6-1	Ratcler	Numeric	2	131–132	
38	6-2	Ratcomp	Numeric	2	133–134	
39	6-3	Ratsupv	Numeric	2	135–136	

Example 13–A (continued)

POSITION	QUESTION #	VARIABLE/FIELD NAME	VARIABLE/FIELD TYPE	FIELD WIDTH	COLUMN BEGIN-END	RESTRICTIONS/COMMENTS
40	6-4	Ratsupp	Numeric	2	137–138	
41	6-5	Ratadms	Numeric	2	139–140	
42	6-6	Ratclnt	Numeric	2	141–142	
43	6-7	Ratinfo	Numeric	2	143–144	
44	6-8	Ratcoop	Numeric	2	145–146	
45	7	Hardcas	Numeric	1	147–147	
46	8	Elsimpv	Numeric	1	148–148	
47	9	Verb9	Alpha	30	149–178	First 30 letters verbatim
48	9-1	Impv1	Numeric	2	179–180	
49	9-2	Impv2	Numeric	2	181–182	
50	9-3	Impv3	Numeric	2	183–184	
51	9-4	Impv4	Numeric	2	185–186	
52	9-5	Impv5	Numeric	2	187–188	
53	10	Verb10	Alpha	30	189–218	First 30 letters verbatim
54	10-1	Comm1	Numeric	2	219–220	
55	10-2	Comm2	Numeric	2	221–222	
56	10-3	Comm3	Numeric	2	223–224	
57	—	Verbcom	Alpha	30	225–254	First 30 letters verbatim
58	—	Icom1	Numeric	2	255–256	
59	—	Icom2	Numeric	2	257–258	
60	—	Icom3	Numeric	2	259–260	
61	—	Intvid	Numeric	4	261–264	
62	—	Intvdat	Date	10	265–274	

Note: This example coordinates with Examples 9-G and 10-A.

Example 13-B

This example illustrates creating a data file layout.

Data Definition for Creating Data File Layout						
Interview of Staff: Data Definition Plan						
POSITION	QUESTION #	VARIABLE/FIELD NAME	VARIABLE/FIELD TYPE	FIELD WIDTH	COLUMN BEGIN-END	RESTRICTIONS/COMMENTS
ID Number		ID	Numeric	4	1–4	leading zeros : KEY
INTERVIEW DATE		DATEYR	Numeric	4	5–8	4-digit year
INTERVIEW DATE		DATEMO	Numeric	2	9–10	
INTERVIEW DATE		DATEDA	Numeric	2	11–12	
	1	LNGWRK	Numeric	1	13–13	1–5, miss 8,9
	2	TYPSRV	Numeric	2 [a]	14–15	01–08, miss 88,99
	2	SRVOTH	String/alpha	20	16–35	Use all lower case, punctuation
	3	TYPPOS	Numeric	2 [a]	36–45	01–07 miss 88,99
	3	POSOTH	String/alpha	20	46–65	Use all lower case
	4	LNGPOS	Numeric	1	66–66	1–6, miss 8,9
	5	RATECOW	Numeric	1	67–67	4–1, miss 8,9
	6	RATESUP	Numeric	1	68–68	4–1, miss 8,9

Example 13–B (continued)

POSITION	QUESTION #	VARIABLE/FIELD NAME	VARIABLE/FIELD TYPE	FIELD WIDTH	COLUMN BEGIN-END	RESTRICTIONS/COMMENTS
	7	RATEADM	Numeric	1	69–69	4–1, miss 8,9
	8	COMMENT	String/alpha	20	70–89	Use all lower case, punctuation[b]
		INTRVWR	String/alpha	40	110–149	First and last names, all lower case
	10	REVIEWCODE	Numeric	3	149-152	Miss = 999

Note: With most software, it is not necessary to write out the beginning and ending column numbers. This example coordinates with Example 9-H.

[a] Two spaces are allowed, although there are fewer than 10 categories. Two spaces make it possible, after all data collection, to code and add the values for SRVOTH or POSOTH to the values for TYPSRV and TYPPOS. This coding procedure is not covered in this guidebook.

[b] If capitals are used in the original data entry, an extra step will be needed to convert all capitals. Capitals can be added in a word processor later.

Example 13–C

This example provides the steps involved in creating a data file layout, with reference to the data provided in Example A, and the steps to be followed when using files based on previously collected and stored data.

Steps for Creating the Data File Layout in Example 13–A

- Establish a code that identifies each review or evaluation. Essential review information includes a coded name that represents the site or location, the date, and sometimes, the review purpose. Give the review or evaluation a variable name and make the necessary specifications. In Example A, the review name is the last variable defined.
- Define the file layout. Separate a sheet of paper into columns. Example A shows how seven columns are typically used to record information. If you wish to document additional information, add columns.
- Label the column heads.
- Write out the information for each field or variable, beginning with the identifying information and including the information shown in Example A and any additional information.
- Follow the same procedures for all of the data collection forms.
- Produce and print a data dictionary for the file. Follow software instructions.
- Determine if any fields should be converted or modified. As an example, a five-space field can be redefined to allow eight spaces, a numeric field can be converted to text, or a field with integers can be converted to one with decimals. Make changes only on a copy of the file. Always keep the original. If you are not experienced in modifying files, get help from an expert.
- Make sure that the identifiers are consistent with the identifiers used for any data collection forms.
- Specify the sort/key.
- Copy the records that represent the population into a file. If necessary, format and mark the key.

After the file has been created, you will be able to generate a data definition automatically from a utility in SPSS and most other statistical or database programs. It is critical that the data entry scheme be planned ahead of time. The instructions for the tasks in Step 13 provide one approach to developing that plan.

Additional Reading and Reference

Rubin, A., & Babbie, E. (2000). *Research methods for social work* (4th ed). Belmont, CA: Wordsworth.

Steps 8–13

Step 8

8-1. Why did the developer-managers decide not to do a comprehensive evaluation?

A comprehensive evaluation studies every aspect of an evaluand, including the department, the unit being evaluated, or the organization, if the whole organization is being evaluated. The Stellar Agency is not studying every aspect of their organization, just those aspects that could be involved in the questions specified in Step 1.

Step 12

12-1. Study the beginning draft plan for data analysis, shown in Table 11. What can Joe, Jean-Claude, and Shaniqua do to determine which resource (telephone, voicemail, or copier) direct service workers perceive as having the worst performance?

Joe, Jean-Claude, and Shaniqua can do the same analysis on voicemail and copier as they did with telephone. They would compute the average number of days that workers reported they had been without the resource. Then, they would compare the average number of days for each resource and determine which had the longest average to identify the worst performance.

Organizing the Review and Evaluation

This part of the guidebook describes and applies the last six steps in preparing to review, evaluate, and study an intervention:

14. Identify the reviewers/data collectors

15. Select the sample and compare sample characteristics to population parameters

16. Assign sample members to data collectors

17. Set up databases

18. Deliver materials and reminders

19. Provide lists to contact individuals at the data collection sites

Step 14:

Identify the Reviewers/Data Collectors

Tasks

- Determine the types of skills required for the data collection. The selected data collection methods will determine which skills are needed (see Appendix A).

- Determine the number of data collectors and how they should be grouped into teams. The number of teams will be determined by the number of sites where data will be collected, the hours of work, and the total days or weeks that can be allocated for the data collection portion of the project. Estimate the time needed to complete data collection.

- Identify the data collection team members and make team assignments.

- Write letters to the individuals whose assistance or cooperation will be needed for a successful data collection effort. These letters should communicate

- the purpose of the data collection;

- who authorized the project (if possible, arrange to have the authorizing individual sign the letter);

- the addresses and telephone numbers of people who can answer questions about the project or verify its validity;

- the reasons for the contact and the type of cooperation or assistance required from the individual;

- a statement regarding the level of confidentiality that can be guaranteed;

- the consequences of not cooperating, if any, stated as gently as possible;

- how the data will be used and how the findings will be reported;

- the names of the data collectors who will be involved in the project; and

- the date and time to expect the contact, or if the communication requests an appointment, the dates and times within which the appointment must occur.

Leaders

Lead Reviewer and Developer-Manager

Skill

- Administering the review process.

Barrier

- Underestimating the number of data collectors or the amount of time needed for data collection.

Binder Task

List the requirements for the review. Next to each item on the list, write the name of a data collector who can meet the requirement.

Application of Tasks

Ruby Brown, the lead reviewer, identified the team from among her trainees. For those she did not select, Ruby indicated whether they simply required extra training and whether they might be assigned to a future project.

Practice Questions

14-1. Review the study process in Step 2, the study expectations in Step 7, the type of study selected in Step 8, and the type of data sought by the surveys in Step 9. Based on this information, write a form letter to the individuals whose assistance will be needed for data collection, as outlined in the Tasks section.

Step 15:

Select the Sample and Compare Sample Characteristics to Population Parameters

Tasks

- Select the sample approximately one week before the review is scheduled to begin. If more than a week will be needed to select the sample and produce and deliver face sheets (see Definition section), allow additional lead time. Allowing more time, however, increases the risk that the population will change between the time of sample selection and the time of the review.

- For population records in a computer file:

- Use a random sample generator available in SPSS and other statistical packages to select the needed sample size, as computed in Step 11.

- Summarize the demographics and key information for the population and the sample.

- Compare sample and population figures. If a difference exists between any statistic and its corresponding population parameter, reselect the sample (that is, return to the same population and reselect, using the same design).

- After final sample selection, convert the records into a format that will allow them to be written to a database file.

- For population records listed on paper:

- Create a database template for entering information about the sampled records.

- Use the sampling interval computed in Step 11 to identify selected records.

- For populations of records in a filing cabinet or drawer:

- For each sampled record, insert a place-keeper and extract the sampled record.

- Copy all information or data indicated by the data collection forms being used for the review.

- Enter the ID number into the appropriate fields of the database.

Definition

Face sheets contain identifying information on the sampled respondent to be included in the review. Sometimes the researcher generates them from the data file when the sample is drawn. The reviewer or data collector uses the face sheet to begin data collection on the respondent.

Leaders

Data Analyst and Developer-Manager

Skills

- Employing research methods and process.

- Designing and selecting samples.

- Employing data analysis process.

Barrier

- Skipping over any step or executing it sloppily.

Binder Task

Review the population table created in Step 11. Add the information in the column for sample information. Cross-reference the notes for Step 15 with the notes for Step 11.

Application of Tasks

Joe sorted the client records so that clients served by each site were in separate files for those sites. He exported the files into spreadsheets that SPSS can use. After Joe opened the Site A file in SPSS, he went to the top menu bar and selected "Data," then chose "Select Data." He selected "Random sample of cases" and clicked the "Sample" button, which opened another menu. In this menu, he selected "Exactly" and typed in 312 to indicate the sample size to be selected, then typed the population size, 600, in the next space. He then clicked "Continue."

Joe used the same procedure for the other three sites and gave the four files to Shaniqua and Jean-Claude. Site A had a sample size of 312, Site B 380, Site C 360, and Site D 346, for a total of 1,398.

Step 16:

Assign Sample Members to Data Collectors

Tasks

- Keep a master list by ID number for each sample member: type of population, name, location, demographic information, and any other pertinent data about the sample member.

- Assign sample members, using their ID numbers, to data collectors. Sample members may be assigned in two ways:

- If all sample members are in a single location, evenly divide the number of sample members among the data collectors. For example, if there are 150 sample members and 10 data collectors, assign each data collector 15 sample members.

- If the individuals to be interviewed or the records to be examined are in several locations, sort sample members geographically, assigning them to data collectors to will minimize travel distances.

- On the master list, enter the name of the data collector for each sample member. Use this form to record the return of forms.

- For each sample member, the quality assurance department should produce a control sheet with the sample member's name and essential demographic data from its databases. Essential demographic data will depend on the purpose of the study and the data the analysis requires.

- Finalize the instructions and QXQ specs described in Step 10. Ensure that the instructions support the accuracy of the data collection process and protect validity.

- Add instructions to the control sheet to guide the data collector on locating the sample member, collecting the data, and correctly and consistently recording all data.

Leaders

Developer-Manager, Reviewer, and Data Analyst.

Skills

- Administering the review process.

- Employing research methods and process.

Barriers

- Large numbers of data collectors. As the number of data collectors increases, greater care is needed to ensure the correct and consistent collection of data.

- Failure to follow protocol for contacting and informing site staff. This failure can result in any number of roadblocks, including absent interviewees, missing records, and improperly drawn samples.

Binder Task

Review the research questions and determine which demographic data are needed. Record these items so they can be entered on the control sheet when the sample is drawn.

Application of Tasks

Using the stratified samples generated in Step 15, Shaniqua and Jean-Claude selected a sample and developed a master list, which they distributed to staff onsite. After consulting with their data collectors, they assigned each case on the sample list to minimize travel distances across the sites. They gave each data collector his or her own review list and a control sheet for each case. In addition, they gave each data collector the QXQ specifications so that he or she could consult the QXQs to be reminded about how an item should be administered. Example 16-A shows a control sheet.

Example 16-A

Data Collection Control Sheet

Data Collector: *Kim Myles*

Site Contact Person: *Brad Lane*

Site/Subsample: A

Sample ID Number: 12

Outcome Codes: 1 = Completed data collection; 2 = pending; 9 = file was not available

Date	Actions Taken	Outcome

Data Collector Comments:

Note: This example coordinates with Examples 9-G and 13-A.

Step 17:

Set Up Databases

Tasks

- SPSS and the layout shown in Example 13-A illustrate the process of setting up the statistical files (see Appendix E). The data file layouts guide the creation of the databases. If the study uses a package other than SPSS, procedures will vary.

Leaders

Data Analyst and Developer-Manager

Skills

- Employing research methods and process.

- Employing data analysis process.

Barrier

- Incorrect specification of widths in the database. Use extreme care when recording and specifying all items, particularly the variable widths.

Binder Task

Begin to assign variable names according to instructions in the statistical package you are using to each item. Ensure that the value for each item, except open-ended questions, has a numeric code.

Application of Tasks

Joe, the data analyst, created the Stellar Agency's databases. All of Stellar Agency's data for the populations are in computer files. After Joe selected a satisfactory sample, he created a copy of the population file with the ID numbers and other fields that would help him analyze systematic differences between responders and nonresponders and between those who refused to answer specific items and those who respond to all items. He will label one blank questionnaire with each sample member's ID number.

Had Stellar Agency's population not been in computer files, he would have noted information on each record (whether an index card, a file in a drawer, or other written document) that could assist him in analyzing differences between responders and nonresponders and between individuals who responded to some questions and individuals who responded to all questions. If the records did not have ID numbers, he would have assigned numbers to records for the population elements to create a key for matching and merging later. The ID numbers on blank questionnaires would have been a field added to records of the databases. Created ID numbers can be coded with information to assist with tracking. For example, an ID number can consist of three parts: the site or region number, the supervisor ID number, and a unique set of numbers that are used consecutively for all sample members.

Step 18:

Deliver Materials and Reminders

Tasks

- Deliver to each data collector the control sheets, the instructions for locating the sample member's record, and the data collection forms. Use an appropriate organizational strategy to ensure the delivery of the proper forms to the correct data collectors. For example, it may be appropriate to use intermediaries to oversee the distribution and receipt of all materials.

- Keep a copy of all materials for monitoring the data collectors' and intermediaries' work.

- Remind contact people at the sites where data collection about delivery of the forms.

Leaders

Developer-Manager and Reviewer

Skills

- Administering the review process.

- Employing data analysis process.

Barrier

- Not getting materials to the data collectors in a timely fashion. To ensure timeliness, use a messenger or express delivery service, if necessary.

Binder Task

Record the dates and locations of deliveries of materials. List the names of the individuals responsible for each step in this process. Make notes of the best ways to contact these individuals.

Application of Tasks

The lead reviewer and the data collectors took the control sheets and the other materials to the sites where they would perform reviews. These materials were left in a locked container for which only the lead reviewer and at least one other data collector had a key. After filing copies of all materials, they made reminder calls to the contact persons at the sites where data collection will occur to ensure that site personnel would remember that the review was to take place.

STEP 19:

STEP 19:

Provide Lists to Contact Individuals at the Data Collection Sites

Tasks

- Two or three days before the review, send the site staff a list of the sample members and a list of the data collectors who will be working at the site. The timing of such communications depends on the proper protocol for the organization.

- Confirm that each site will provide space that the data collectors can use to review records and, if planned, conduct interviews.

Leaders

Data Analyst and Developer-Manager

Skill

- Administering the review process.

Barrier

- Failure to communicate with data collectors or site staff.

Binder Task

Keep notes of your communications with data collectors and site staff. Date everything you do.

Application of Tasks

Ruby, the lead reviewer, sent the staff at the sites a list of the sampled records, the individuals selected for interviews, and a list of the data collectors scheduled to work at each site. She stated her understanding that the team would have private space for reviewing records and conducting interviews and that the space could be locked.

PRACTICE QUESTION AND ANSWER

Steps 14–18

Step 14

14-1. Review the study process in Step 2, the study expectations in Step 7, the type of study selected in Step 8, and the type of data sought by the surveys in Step 9. Based on this information, write a form letter to the individuals whose assistance will be needed for data collection, as outlined in the Tasks section.

This is a sample letter based on Stellar Agency's study.

Dear _____ ,

The Stellar Agency is conducting a quality assurance survey on its service delivery, authorized by the leadership council. Based on your contact with the agency, you have been selected for participation in this study.

My name is (Name), and I am a data collector for this study. I can be reached at (Phone number). I will be calling you soon to schedule a time to meet. The dates available are:

(Dates)

Participation in this study is voluntary. This means that if you decide not to participate, your relationship with the agency or your caseworker will not be affected.

Attached are a copy of the agency's confidentiality policy, which will be observed through-out the study, and an explanation of how the data from this study will be handled to keep confidentiality. These will be explained to you by one of our data collectors. You may also call the staff in charge of this study:

- *Shaniqua Grabowsky at (Phone number) in Region 1*

- *Jean-Claude Washington at (Phone number) in Region 2*

Thank you for your important contribution to this study!

Sincerely,

(Signature)

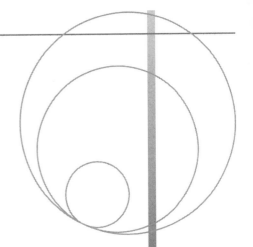

Summary of Stellar Agency's Analysis

This guidebook discusses how to plan and prepare a quality analysis. In presenting this process, examples for varying situations were given. This summary brings together the 19 steps you have learned and shows how Stellar Agency has used them to prepare a quality evaluation. Using this summary to take stock of your own evaluation ensures that nothing has been missed and that your agency is prepared to implement the project.

The goal of Steps 1 through 7 is to define the issues and participants of the project, whether it is a review or an evaluation. The dependent concept, or main phenomenon, that the Stellar Agency will study is the amount of time that direct service workers have to wait for critical resources so that they can continue their work with clients. They want to know if direct service workers wait a long time, so that the organization can seek solutions to shorten waiting periods and improve service.

This dependent concept, although good, is very broad. There are many things that could be critical resources for direct service workers, and the leadership team needs to know not only whether direct service workers wait a long time, but also which resources they wait for the longest. Stellar Agency does not want to waste time and effort trying to speed up delivery on all resources, when it might need to only concentrate on specific ones. To this end, Shaniqua operationalized the dependent concept, or separated the dependent concept into discrete, measurable core variables.

She then defined each component operationally. Operational definitions contain all of the information necessary for measurement. She operationalized resources into the following:

- office/clerical support,

- computer applications,

- supervisor input,

- office supplies,

- administrative input,

- resource for client,

- client information, and

- client cooperation.

To reemphasize that operationalization is guided by the needs of the specific study, note that all of the dependent variable components might have been broken into smaller components. For example, specific types of office supplies could have been specified (e.g., pens, paper clips), but this study did not require that much detail.

Time waited for resources was defined as when the resource was available to the responding direct service worker:

- at the time first needed,

- a few days after first needed,

- a week or two after first needed,

- longer than two weeks after first needed, or

- never received at all.

Next, Stellar Agency identified the independent and control variables. The independent variables are those things that the Stellar Agency can control and, if problematic, could improve. The independent concept for this project is the types and adequacy of services or equipment. This independent concept was operationalized into the following variables:

- available telephone,

- effective message receipt,

- available copier or copy service,

- available supplies,

- number of full-time-equivalent staff, and

- number of staff vacancies that to be filled.

The control variables, over which the Stellar Agency has no control but which could enhance or impede direct service workers' progress with clients, are

- severity of the family's problems,

- size of the family,

- number of staff vacancies that were eliminated, and

- average caseload size.

After Shaniqua defined the variables, the Stellar Agency quality assurance team identified all of the stakeholders relevant for this project, to ensure that the appropriate outcomes were being produced. They then identified the relevant populations from which to select samples; determined who would be reading and using findings, so that reports could be tailored for the primary users; and explored different data collection methods in relation to the project to find the most appropriate method for their project.

Steps 8 through 11 covered specific review and evaluation design issues. After making several important decisions in Step 8, Stellar Agency undertook the task of developing or adapting data collection items and forms. Next, the developer-managers developed materials for training the data collectors. It might seem out of order to develop QXQ specifications and data collection guides now, however, developers should design them as soon as they believe their data collection forms are final, and before going to the expense of having them duplicated. Designing these materials affords the developers an opportunity to scrutinize the data collection forms from a different perspective, which allows a chance to catch problems missed in previous examinations.

Then, it was time to design the samples, compute sample sizes, draft the preliminary data analysis plan, and design database layouts. Tackling these tasks forces you to look at the data collection forms from yet another viewpoint and allows you to re-examine them before duplication. The most important reason, however, is that all of these things must work together, so developing and revising them together is essential. The database layouts must reflect forms precisely. The data collection forms contain representations of the variables. It is easy now to jot down how you will summarize and analyze variables together.

Steps 14 through 19 allow Stellar Agency organize its review: Ruby identified the team of data collectors. Jean-Claude and Shaniqua selected the sample just before Ruby trained the data collectors, to reduce the chances that the population Stellar Agency is studying will have changed greatly. They then assigned each data collector a set of sample members, sent letters to the potential respondents, trained the data collectors, and had Joe create and test their database. They rechecked everything to ensure that they are ready to perform the evaluation. Once all this has been done, and you are sure everything is in place, you are ready to begin collecting data.

Additional Readings

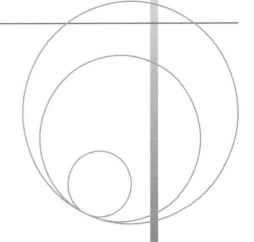

Araoz, D. L., & Sutton, W. S. (1994). *Reengineering yourself: A blueprint for personal success in the new corporate culture.* Holbrook, MA: Bob Adams.

Argyris, C. (1990). *Overcoming organizational defenses: Facilitating organizational learning.* Englewood Cliffs, CA: Prentice Hall.

Ashford, S. J. (1988). Individual strategies for coping with stress during organizational transitions. *Journal of Applied Behavioral Science, 24,* 19–36.

Babbie, E. R. (1998). *The basics of social research.* Florence, KY: Thompson Learning.

Baker, H. K., & Holmberg, S. R. (1982). Stepping up to supervision: Coping with change. *Supervisory Management, 27*(3), 21–27.

Becker, H. (1999). *Tricks of the trade: How to think about your research while you are doing it.* Chicago: University of Chicago Press.

Beckhard, R., & Prichard, W. (1992). *Changing the essence: The art of changing and leading fundamental change in organizations.* San Francisco: Jossey-Bass.

Belasco, J. A., & Stayer, R. C. (1993). *Flight of the buffalo: Soaring to excellence, learning to let employees lead.* New York: Warner Books.

Bennett, M. J. (1998). Intercultural communication: A current perspective. In M. J. Bennet (Ed.), *Basic concepts of intercultural communication: Selected readings* (p. 1–34). Yarmouth, ME: International Press.

Bennis, W. (1997). *Organizing genius: The secrets of creative collaborations.* Reading, MA: Addison-Wesley.

Bertcher, H. J., & Maple, F. F. (1996). *Creating groups.* Thousand Oaks, CA: Sage.

Blalock, H. M. (1979). *Social statistics.* New York: McGraw-Hill.

Blazey, M. (1998). *Insights to performance excellence 1998: An insider look at the Baldrige Award criteria.* Milwaukee, WI: Quality Press.

Block, P. (1995). *Stewardship: Choosing service over self-interest.* San Francisco: Berrett-Koehler.

Briscoe, D. R. (1977). Toward an individual strategy for coping with change. *Personnel Administrator, 22*(7), 45–48.

Brook, A. (1978). Coping with the stress of change. *Management International Review, 18*(3), 9–15.

Burleson, B. R, Albrecht, T. L., & Sarason, I. G. (1994). *Communication of social support: Messages, interactions, relationships, and community.* Thousand Oaks, CA: Sage.

Champy, J. (1995). *Reengineering management: The mandate for new leadership.* New York: Harper Business.

Cohen, M. D., & Axelrod, R. (1984). Coping with complexity: The adaptive value of changing utility. *American Economic Review, 74*(1), 30–42.

Connor, D. R. (1995). *Managing at the speed of change: How resilient managers succeed and prosper where others fail.* New York: Villard Books.

Cook, T. D., & Campbell, D. T. (1979). *Quasi-experimentation: Design and analysis issues for field settings.* Boston: Houghton Mifflin.

ADDITIONAL READINGS

Dartington Social Research Unit. (1995). *Child protection: Messages from research*. London: HMSO.

Davenport, T. H. (1993). *Process innovation: Reengineering work through information technology*. Boston: Harvard Business School Press.

deBono, E. (1994). *deBono's thinking course* (Rev. ed.). New York: Facts on File.

Deep, S., & Sussman, L. (1995). *Smart moves for people in charge: 130 checklists to help you be a better leader*. Reading, MA: Addison-Wesley.

DePorter, B., & McPhee, D. (1996). *Quantum learning*. New York: DTP Trade Paperbacks.

DePree, M. (1989). *Leadership is an art*. New York: Dell.

DePree, M. (1992). *Leadership jazz: The art of conducting leadership through leadership, followership, teamwork, touch, voice*. New York: Dell.

Fournies, F. F. (2000). *Coaching for improved work performance*. New York: McGraw-Hill.

Gilmore, T. N., Shea, G. P., & Useem, M. (1997). Side effects of corporate cultural transformations. *Journal of Applied Behavioral Science, 33*, 174–189.

Hammer, M. (1996). *Beyond reengineering*. New York: Harper Collins.

Hammer, M., & Stanton, S. A. (1995). *The reengineering revolution: A handbook*. New York: Harper Collins.

Inman-Ebel, B. (1999). *Talk is not cheap! Saving the high cost of misunderstanding at work and home*. Austin, TX: Bard Press.

Institute for Quality Assurance. (2000). *The quality body of knowledge for professionals and the eight knowledge sections*. Retrieved from http://www.iqa.org.

Kolb, D. A. (1984). *Experiential learning: Experience as the source of learning and development*. Upper Saddle River, NJ: Prentice Hall PTR.

Kolb, D. A. (1999). *Learning style inventory*. Boston: TRG Hay/McBer.

Kotter, J. P. (1996). *Leading change*. Boston: Harvard Business School Press.

Lincoln Foundation for Excellence. (1998). *The 1998 Lincoln Award criteria*. Chicago: Author.

Maccoby, M. (1997). Making many Penny's: A new management challenge. *Research-Technology Management, 40*(1), 56–58.

Martin, J. (1995). *The great transition: Using the seven disciplines of enterprise engineering to align people, technology, and strategy*. New York: American Management Association.

McPhee, D. (1996). *Limitless learning*. Tuscon, AZ: Zephyr Press.

Mears, P. (1995). *Quality improvement tools and techniques*. New York: McGraw-Hill.

Naisbitt, J. (1982). *Megatrends: Ten new directions transforming our lives*. New York: Warner Books.

Naisbitt, J. (1994). *Global paradox*. New York: Avon Books.

Naisbitt, J., & Aburdene, P. (1990). *Megatrends 2000: New directions for tomorrow*. New York: Avon Books.

Nanus, B. (1992). *Visionary leadership: Creating a compelling sense of direction for your organization*. San Francisco: Jossey-Bass.

Peak, M. (1997). Go Corporate U! *Management Review, 86*(2), 33–37.

Peters, T. (1992). *Liberation management: Necessary disorganization for the nanosecond nineties*. New York: Fawcett Columbine.

Quinn, R. (1996). *Deep change: Discovering the leader within*. San Francisco: Jossey-Bass.

Rossett, A. (1997). That was a great class, but.... *Training & Development, 51*(7), 18–24.

Rossi, P. H., & Freeman, H. E. (1993). *Evaluation: A systematic approach*. Newbury Park, CA: Sage.

Rubin, A., & Babbie, E. R. (2000). *Research methods for social work* (4th ed.). Belmont, CA: Wordsworth.

Scheaffer, R. L., Mendenhall, W., & Ott, L. (1979). *Elementary survey sampling* (2nd ed.). North Scituate, MA: Duxbury Press.

Schwartz, P. (1991). *The art of the long view: Planning for the future in an uncertain world.* New York: Currency Doubleday.

Senge, P. M. (1990). *The fifth discipline: The art and practice of the learning organization.* New York: Currency Doubleday.

Senge, P. M. (Ed.), Kleiner, A. (Ed.), Roberts, C., Ross, R., & Smith, B. (1994). *The fifth discipline field book: Strategies and tools for building a learning organization.* New York: Currency.

Sluyter, G. V. (1998). *Improving organizational performance: A practical guidebook for the human services field.* Thousand Oaks, CA: Sage.

Stewart, T. A. (1997). *Intellectual capital: The new wealth of organizations.* New York: Currency Doubleday.

Sudman, S., & Bradburn, N. (1990). *Asking questions: A practical guide to questionnaire design.* San Francisco: Jossey-Bass.

Tropman, J. E. (1995). *Effective meetings: Improving group decision-making.* Thousand Oaks, CA: Sage.

Vaill, P. B. (1996). *Learning as a way of being: Strategies for survival in a world of permanent white water.* San Francisco: Jossey-Bass.

Wheeler, D. J. (1993). *Understanding variation: The key to managing chaos.* Knoxville, TN: SPC Press.

Winter, R. P., Sarros, J. C., & Tanewski, G. A. (Jan 1997). Reframing managers' control orientations and practices: A proposed organizational learning framework. *International Journal of Organizational Analysis, 5*(1), 9–24.

Zirps, F. A. (1997). *Doing it right the first time: A model of quality improvement for human service agencies.* Orlando, FL: Florida Institute of Quality Improvement.

APPENDIX A

Quality Skills and Abilities

RESEARCH SKILL	ABILITIES
Employing evaluation methods	a. Knowledge of the evaluation process and the tasks involved in each step. (See Appendix F.) b. Knowledge regarding the implementation of each evaluation step and the circumstances under which assistance should be obtained for tasks that cannot be performed alone. c. Ability to identify resources for data in addition to case records.
Employing research methods and process	a. Knowledge of the research process and the tasks involved in each step.
Performing a stakeholder analysis	a. Using a problem-solving process.
Selecting characteristics to be included and excluded in the population definition	a. Using a problem-solving process.
Designing and selecting samples	a. Ability to select the appropriate design for the defined population and the specific situation. b. Understanding the implications of unbalanced returns, when it is appropriate to use weighting or other techniques to adjust for unbalanced returns, and how to make adjustments. c. Ability to determine whether a sample is similar to the population.
Developing and designing questionnaires	a. Ability to define the concepts and behavioral phenomena that represent the policy, practice, or research questions common to quality projects. b. Ability to operationalize concepts the data will measure. c. Ability to write unambiguous data collection form items. d. Ability to write items so they correspond to the unit of analysis. e. Ability to select the appropriate level of measurement for each item. f. Ability to develop response categories in appropriate formats for item stems and scales. g. Ability to order the items appropriately for the reviewers.

(continued)

MANAGMENT SKILL	ABILITIES
Employing the data analysis process	a. Ability to use probability statements from findings to construct analysis tools, such as tables and graphs.
	b. Ability to use these tools to identify additional statistical procedures to answer questions that arise from working with the data and reaching preliminary conclusions.
	c. Ability to conclude that all questions that can be answered have been answered, that is, the ability to understand when the limits of the data have been reached.
	d. Ability to use findings from the literature or unpublished studies to speculate appropriately on issues related to the current study's findings.
Administering the review process	a. Knowledge of the complete quality assurance and quality improvement processes and the tasks involved.
	b. Ability to perform quality assurance, improvement and management tasks as a member of a review team.
	c. Ability to identify data sources for quality evaluation judgments.
Employing problem-solving methods	a. Ability to define a problem completely to allow identification of essential components.
	b. Ability to identify activities that can result in the elimination of the problem and specify short- and long-term expected outcomes for the activities.
	c. Ability to develop strategies for implementing the activities.
	d. Ability to use the method to monitor progress in resolving the problem.
	e. Ability to modify the model or plan to improve its effectiveness.
Speaking effectively at meetings, during group processes, and in one-on-one exchanges	a. Knowledge of various communication styles and how those styles differ.
	b. Ability to speak clearly and concisely to maximize understanding.
	c. Ability to consistently apply the principles of intercultural communication.

APPENDIX B

Validity and the
Validity Review Process

Validity is the best available approximation to the truth in propositions about phenomena being studied. The modifiers *approximately* and *tentatively* are understood as prefaces to *validity* and *invalidity*, because researchers can never know what is true (Cook & Campbell, 1979). Unlike reliability, researchers have no objective test for validity that will produce a coefficient. Rather, the developer-manager must use a logical process to protect the four types of validity and use rational discussions to discount competing hypotheses, that is, phenomena other than the intended ones that could account for the obtained results. Cook and Campbell (1979) discussed four types of validity. After this section describes the validity types, it presents a process that can be used to protect the validity of review or evaluation findings.

Validity Types

This section presents validity types in the order in which each type of validity is a major consideration. Progressing from one validity type to another, however, does not mean that the developer-manager can ignore validity types handled previously. All research steps are integrally linked, and research is a recursive process. The developer-manager must always remain alert concerning how decisions made at later steps may require a return to earlier step to make modifications. The descriptions here are not intended to be complete discussions of validity. It is highly recommended that readers consult specialized sources, such as *Quasi-experimentation: Design and Analysis Issues for Field Settings* (Cook & Campbell, 1979), for a complete discussion.

Construct Validity

This type of validity is most threatened when developer-managers are developing, defining, and operationalizing concepts. The heart of construct validity is the avoidance of all fuzziness in meaning. Unintended phenomena should not appear in definitions, indicators, or conclusions. To avoid this problem, developers should create clear definitions for all phenomena that might become study variables. Construct validity also is an issue when selecting or developing items used to collect data. Cook and Campbell (1979) identified nine common threats to construct validity:

- inadequate preoperational explication of constructs,
- mono-operation bias,
- mono-method bias,
- hypothesis-guessing when respondents persist in trying to discover what the experimenters want to learn from the research within experimental conditions,
- evaluation apprehension,
- experimental expectancies,

- confounding constructs and levels of constructs,

- interaction of different treatments, and

- interaction of testing and treatment.

External Validity

External validity means that the researcher can generalize results and conclusions to a defined population, whether composed of people, settings, or times. This type of validity is at issue when the developer-manager is defining populations and designing samples, and after he or she completes data collection. External validity can be addressed by using probability sampling, with selection probability set according to each strata's proportion in the population. After data collection, the issue is whether the completion rates within each sample stratum correspond to proportional sizes in the population. An appropriate random sample of the population can address part of this issue by taking into consideration hetereogeneity, the probability of selecting rare elements of interest, and other issues. Other issues may arise, however, in connection with the completion rate and the breakdown of the final, observed sample. An overall completion rate of 90%, for example, will not help if the breakdown among subgroups is skewed when compared with the breakdown in the population, such as when 60% of completed cases come from one of four subpopulations and each subpopulation comprises 25% of the larger population. Cook and Campbell (1979) identified three common threats to external validity:

- interaction of selection and treatment,

- interaction of setting and treatment, and

- interaction of history and treatment.

Internal Validity

This type of validity is the prime concern at every step of the research process. Internal validity indicates that the results approximate the reality. All aspects of research planning, design, and implementation affect internal validity, and as a result, the developer-manager must be vigilant in protecting internal validity. Internal validity is compromised if one important variable is omitted from the study, defined inadequately, or represented inadequately by an indicator. Internal validity also can be compromised during the data collection period by a number of potential threats. Cook and Campbell (1979) identified nine common threats to internal validity:

- history,

- maturation of the respondent,

- testing,

- instrumentation,

- statistical regression,

- selection,

- mortality when different kinds of persons drop out of a treatment group during the experiment,

- various interactions with selection (selection-maturation, selection-history, selection-instrumentation), and

- ambiguity about the direction of causal influence.

Statistical Conclusion Validity

This type of validity can be threatened during the data analysis process. As with other types of validity (and, indeed, with the entire research process), however, decisions made prior to analyzing data can affect statistical conclusion validity. Two types of power are at issue. One is whether the sample size will allow detecting the true magnitude of relationships. The second is whether the interval constructed around results at a 95% confidence level is narrow enough to be meaningful (Blalock, 1979). Cook and Campbell (1979) identified seven common threats to statistical conclusion validity:

- low statistical power,

- violated assumptions of statistical tests,

- fishing and the error rate problem,

- reliability of measures,

- reliability of treatment implementation,

- random irrelevancies in the experimental setting, and

- random hetereogeneity of respondents.

The Validity Process

This process can protect the validity of review or evaluation findings:

- Use a design to protect validity where it is most important. Think through every phase of the process, including defining the concepts, constructs, and population; designing the sample, developing selection strategies and sampling elements from the population; collecting data and monitoring fidelity of the implementation process, compared with the plans; reaching conclusions based on analyses; and and formulating statements for disseminating findings. Build as many defenses as possible for each validity type.

- Determine what could go wrong with the plan process (and something will!) and develop as many alternate plans as necessary.

- Develop items and procedures to collect data on potential problems that cannot be prevented. Should one of these problems occur, be prepared to assess its effect on the project.

- Keep watch during data collection to prevent compromising results.

- Review the project and analyze it to determine whether validity has been compromised.

- If an unanticipated threat to validity occurs, decide whether it is actually might compromise validity. Then, make notations of your judgments.

- Critically evaluate whether findings, assertions, and speculations can be accepted or should be rejected because the findings may be the result of phenomena other than the actions taken to achieve intended outcomes. For each competing hypothesis, determine whether the competing hypothesis can be discounted. If it cannot be discounted, develop a caveat statement. For example, "Caution should be observed regarding any findings based on results for strata with completion rates of less than 60%."

The Research Process

1. Establish recordkeeping and process monitoring systems. Use a loose-leaf notebook to maintain documentation on the research. Use a computer database to manage the population, sample, and survey data. It is extremely important to keep good records on the status of questionnaires, including return rates, and other data collection instruments.

   Develop a work plan. Choose manual or computerized methods to update the work plan, and update it frequently.

2. Identify and define the problem conceptually. The definition will be refined several times during the process. Divide the problem into its basic parts, so that the researcher can develop effective conceptual definitions of the major variables.

   Identify the main issue to be studied. Usually, the main issue is the dependent variable. Describe the main issue in enough detail that someone unconnected with the research can understand it. Recruit one or two colleagues to be "reactors," and assist with any clarification that is needed.

   Identify the other issues that play a role in this issue, such as policy changes, lack of resources, and workload issues.

3. Refine the definition of the problem. Consult the literature, others who have studied the problem or related topics, professionals who practice in fields related to the topic, and others with knowledge of key aspects of the problem.

   Based on the information obtained:

   - verify that an answer to the problem is not already available;

   - identify phenomena that are suspected, generally accepted, or known to be related to the problem in some way; and

   - record the definitions that other people developed for these phenomena in the various sources that you consulted.

   - Based on this information, revise or refine all conceptual definitions of the major variables to be studied.

4. Formulate the research questions.

5. Describe how the other variables seem to be related to the dependent variables after consulting relevant sources. For example, lack of sufficient staff can affect timely sibling visitation.

6. Organize the knowledge from other sources into a system of related statements or questions. Taking into account all of the statements and questions, reformulate the research questions from Step 4 into a small number of overarching research questions. A limit of three overarching questions is best, but the study can have up to five. Keep rewriting the research questions until they are as clear as they can be.

7. **Begin to design the research.**

   Determine the unit of analysis for the study. The unit of analysis will correspond directly to the population unit. It may result in a problem with ecological fallacy, but this outcome is unlikely.

   Carefully consider the research questions, the expanse of empirical information available on the research topic, and the degree to which available findings help illuminate the issues. Decide, based on this information, whether the project is primarily exploratory, descriptive, explanatory, or hypothesis testing. (Evaluation research is hypothesis testing, that is, it answers the question, "Did this intervention cause the client's condition to improve?").

   Develop operational definitions for all phenomena in the study: dependent, independent, and control. The operational definitions point directly to what is observed when a variable is present. These definitions, in all likelihood, will be refined several times before they are usable.

   Consider research designs. One-group designs include cross-sectional, statistical, pretest and posttest designs. Multiple-group designs include experiment and control (matched and random selections) and multiple comparison groups. Consider the strengths and weaknesses of each relevant design.

   Select the design that provides most validity protection overall and leaves only vulnerable areas that are relatively easy to defend. Develop defenses for vulnerable areas or ways to demonstrate internal validity.

   Refine the research questions again. Refine all definitions again.

   At this point, the researcher will not have selected the methodology, whether qualitative, quantitative, or grounded theory. Her or she must develop the project more fully to make this decision. The methodology will flow from a range of considerations, including the issue being studied, how much empirical knowledge is available on the issue, the research design, and the extent to which he or she can refine definitions. It is not appropriate to start with a desired methodology and then design the research.

8. **Define the population and sample.**

   Considering the unit of analysis (for example, an individual, a group, or an organization) from the previous step, define the parameters for these units. Consider the following:

   - What are the geographical boundaries?

   - How far back in time will the study reach?

   - If individuals are the units of analysis, should both males and females be included? Should the individuals be within a particular age range? Do race or ethnicity matter?

If the research examines the behaviors of a single individual over time (known as a single-subject design), the unit of analysis is not the person, but the behavior.

If the unit of analysis is larger, such as a group (in situations in which individual scores are unavailable or the researcher averages or aggregates individual scores to represent the group score), beware of the ecological fallacy. Ecological fallacy is drawing conclusions from analyses with aggregated units and applying them to the individuals that comprise the aggregated groups.

After selecting the appropriate unit of analysis, define the population, even if the research will not take a sample. Ensure that the population definition makes it possible to answer the research questions.

Obtain measures of the population's demographics and other important characteristics. This information is necessary for designing the sample and evaluating the sample selection. If only estimates are available, obtain the best available data from literature, census figures, or unpublished statistics.

How does this fit in the discussion of the definition of the population? If a baseline is required, determine requirements for baseline adequacy. A baseline is needed if you want to compare your population before and after an intervention to determine amount of change.

After defining the population and deciding to select a sample from the population, the following considerations should be made.

If sampling is to be done, determine which sample designs are appropriate for the project. Consider issues such as

- cost;
- difficulty of locating qualified sample elements, such as families with exactly two children; and
- whether conclusions about a small subpopulation are needed, such as conclusions about services to males over 80 years.

Design the sample:

- Determine the required sample size, using formulas for computing minimum sample size.
- Develop a strategy to select the sample.
- Develop a plan for gaining sample members' cooperation and for documenting willingness to participate.
- Select the sample and obtain measures of the sample's important characteristics. Compare them with population characteristics.
- If required, determine the size of the baseline sample and length of time required for an adequate baseline.

---

9. **Draft the introduction of the report and select a methodological approach for the project. Write a description of the approach you have selected.**

---

10. **Develop the measures. Consider the kinds of findings needed to answer the research questions. How detailed must the findings be? What type of information is required? Clarifying these issues will assist in decisions regarding**

the measurement of variables, the level of measurement of the variables, and the types of statistical procedures that will be appropriate for the data. Reexamine and refine operational definitions, and develop the preliminary data analysis plan.

11. **Develop the data collection instruments. Develop and/or identify data collection instruments. The following situations describe the steps to take in different situations involving data collection instruments.**

Situation 1: Development of a New Questionnaire or Instrument

- Use the operational definitions and information about level of measurement to develop items for data collection tools. Consider which items can be precoded and which items must be open-ended. Think about the possible range of responses for open-ended items.

- Format items into a questionnaire or instrument.

- Pilot test the items and the format. Revise the items and format based on pilot test results.

- If a baseline is required, collect baseline data and establish interrater reliability. Interrater reliability means that two or more independent observers or judges agree that an item is measuring what it is suppose to be measuring.

Situation 2: Selection of Items in a Previously Used or Tested Questionnaire or Instrument

- Use the operational definitions and level of measurement information to set criteria for selecting or rejecting items or groups of items to be used together.

- Modify items for this study's use. Attempt to minimize modifications.

- Format items into a questionnaire or instrument.

- Pilot test the modified items and the format. Revise based on pilot test results.

- If a baseline is required, collect baseline data and establish interrater reliability.

Situation 3: Use of Secondary Data

- Obtain a copy of the questionnaire or instrument used to collect the original data.

- Use the operational definitions to set criteria for selecting or rejecting data elements.

12. **Design data analysis plan. Review the research questions. What information do the questions require? Determine which variables must be analyzed to obtain the information that is needed to answer the questions. Will the variables need to be examined alone or with other study variables? What types of statistical procedures must be used to obtain the answers?**

13. **Develop other documents needed for the project, including advance letters, consent forms, permission forms, assent forms, statements of confidentiality, instruction manuals, and staff schedules. Be sure to secure supplies, including letterhead paper and envelopes. Ensure that the record keeping and monitoring systems are operational for the project.**

14. Send any advance materials to the data collectors and sites where the data collection is to occur. Document the mailing of all materials and the return of all consent forms.

15. Once consent has been obtained (or if consent is not required), begin collecting data. Conduct field observations, mail questionnaires, begin telephone contacts, initiate interviewing, and conduct focus groups.

16. Edit, code, and process the data.

17. Modify the data analysis plan.

18. Finish the first draft of the methodology description.

19. Analyze the data according to the plan.

20. Record each result as produced.

21. Produce tables and charts to summarize the results. These will be the tools that help analyze and write up the findings.

22. Finalize the introduction and methodology description.

23. Edit and rewrite the project document. Obtain an editor , if possible, to review your document.

24. Write the findings of the data analysis using the tables and charts produced in the earlier step.

25. Polish the document. Edit, rewrite, proof, and forward to the editor.

26. Publish, present, and disseminate findings.

APPENDIX D

Stellar Agency's Organizational Demographics

	Site A	Site B	Site C	Site D	Total
Employees	53	96	80	71	300
Direct service workers	40	72	60	53	225
Clerical workers	8	14	12	11	45
Client					
Organizations	2	20	25	18	65
Individuals					
Contract and noncontract (by age)					
0–4	149	378	270	239	1,036
5–12	151	378	270	245	1,044
12–17	150	270	270	230	920
18–64	100	32	45	42	219
65+	50	22	45	39	156
Total	600	1,080	900	795	3,375
Contract (by age)					
0–4	75	302	176	155	708
5–12	75	300	172	159	706
12–17	111	262	180	150	703
18–64	0	32	31	27	90
65+	20	22	28	25	95
Total	281	918	587	516	2,302
Noncontract (by age)					
0–4	74	76	94	84	328
5–12	76	78	98	86	338
12–17	39	8	90	80	217
18–64	100	0	14	15	129
65+	30	0	17	14	61
Total	319	162	313	279	1,073
Families					
Contract	40	204	130	101	475
Noncontract	110	12	50	58	230
Total	150	216	180	159	705
Groups					
Contract/grant	6	105	70	55	236
Private pay	68	20	40	20	148
Community compact	1	30	20	25	76
Total	75	155	130	100	460

APPENDIX E

SPSS Database Illustrations

1. When SPSS is opened, the application is in Data View. Move the cursor to the bottom-left of the screen and switch to Variable View.

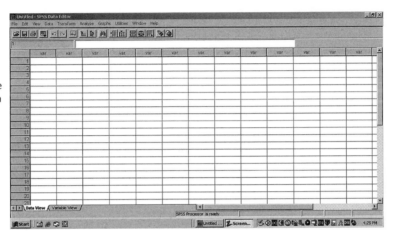

2. In Variable View, the properties of SPSS variables are displayed across the top of the table. Click "Name" on the first row to begin defining the variables.

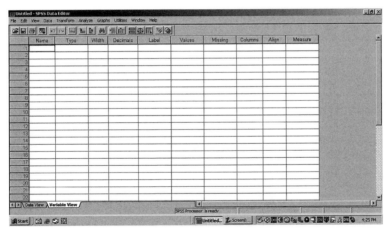

3. Type in the name of the first variable, id. SPSS stores all variable names in lower case; caps lock has no effect on variable names. Enter information to define id. To change defaults, click on the field to enter information or display a list of options. The next frame is an example.

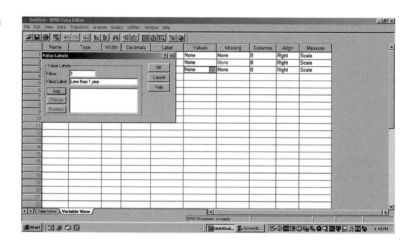

4. Entering labels for values.

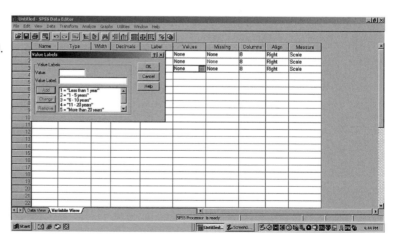

5. Display of labels for values that have been added.

6. Click "Measure" to change the variable's level of measurement, as shown in this frame and the next.

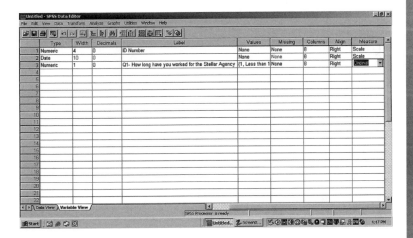

7. Display of changing the level of measurement.

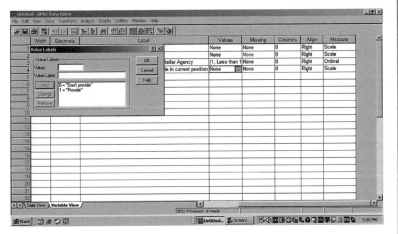

8. This item allows respondents to check all that apply, so each selection is defined as a separate variable. If checked, the respondent provides the service, defined as "1." If not checked, the respondent does not provide it, defined as "0."

9. In most instances, if you make a mistake, click twice in the field and edit to correct. In the example, change Provice to Provide.

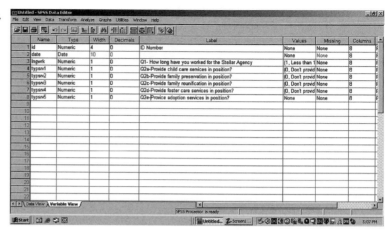

10. When several variables have the same values, copy and paste, as shown in this frame and the next.

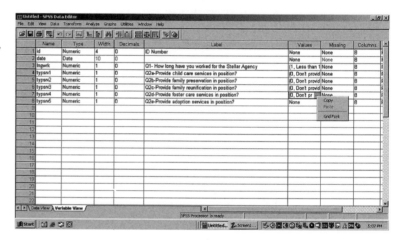

11. Display of value labels being pasted.

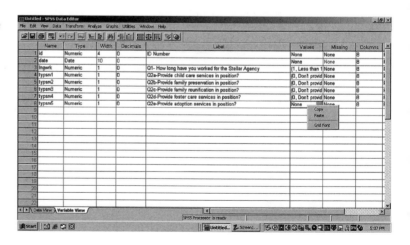

12. Display of changing the variable type to "String," which is the SPSS term for text variables

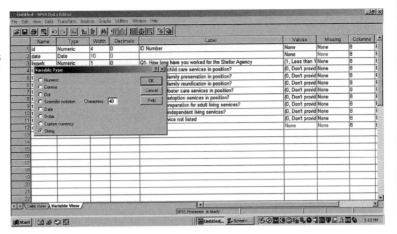

13. This frame displays properties of the variables that have been defined so far. The variable list after the current variable, otherpos, has been completely defined.

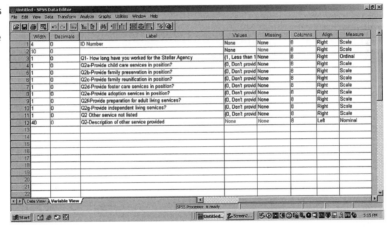

14. Defining level of measurement for lngpos.

15. Defining values
for co_rate.

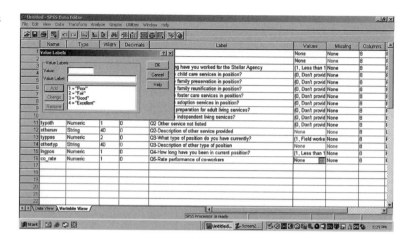

16. Changing
variable type
for comments.
Respondents'
comments will
be paraphrased
in SPSS and
entered verbatim
in Word or
WordPerfect.

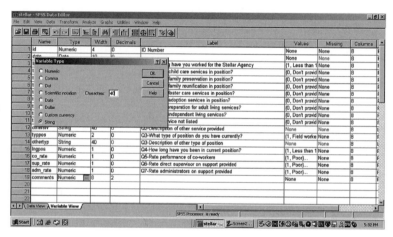

17. Display of most
variables after
the definition
process has been
completed; the
first three (id,
date, lngwrk)
do not show
due to scrolling.

18. After the definition process has been completed, switch back to Data View to begin entering data.

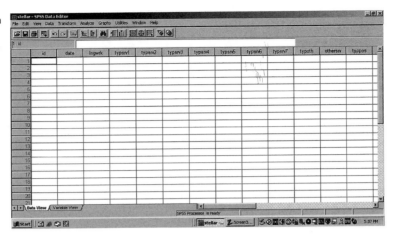

Steps in Conducting Effective Evaluations

ACTION	DECISIONS TO MAKE AFTER TAKING ACTION
Examine the evaluand (the program, project, service, or situation to be evaluated using a problem-solving and analysis tool such as Logic Modeling) to • determine the evaluand's expected outcomes; • confirm that the evaluand's functions are consistent with its expected outcomes; • ascertain assumptions around which the evaluand is designed, including known or suspected barriers; and • describe the tasks performed with the intentions of achieving expected outcomes.	Identify anticipated outcomes as one or more criteria of merit for the evaluand. These outcomes often are specified in official documents.
Determine if the evaluand should be judged on criteria of merit in addition to the outcomes Step 1 identified. Known, planned merit criteria may not in official documents, and criteria may have changed after implementation without official documents being updated.	Decide the order of importance of the merit criteria.
For each merit criterion, determine best indicators of success and failure. What will allow a conclusion that the evaluand is successful? What will prevent judgments of success?	Within each criterion, order indicators in terms of their effectiveness for discriminating between success and failure. Values or conditions that indicate success must always be more positive than what is likely to occur without the evaluand's intervention. Estimate values or describe scenarios as if evaluand's activities or services did not exist. This process will require a determination of any other systematic or exogenous phenomena that could affect the evaluand's status on the problem, positively or negatively. The values for "what probably would occur" should take these systematic factors into account.
Design the evaluation.	Review the type of evaluation. Create protocols: the evaluation design, data collection methods, instruments and forms, procedures, and likely analysis methods for measuring each on the criteria. Depending on labor intensity and expense, consider whether some criteria should be eliminated. Keep detailed explanations of the reasons any criteria were not used.

ACTION	DECISIONS TO MAKE AFTER TAKING ACTION
Collect and analyze the data.	Create the databases for tracking returned instruments and analyzing the data. Enter and keep track of completed instruments or forms.
Compute completion rates; compare them with beginning samples.	Decide if the final sample is representative of the beginning sample. If not, determine whether the final sample can be weighted to bring it into line with the beginning sample. If weighting is not an option, make a note to consider this issue when reaching conclusions and making recommendations.
Compare actual outcomes with expected outcomes to learn whether results are the same, better, or worse than expected.	If the results are the same or better, evidence supports an initial conclusion that the intervention was successful. If the results are worse, evidence supports the conclusion that the intervention failed. The evaluator should look for reasons for failure so that program planners and staff can modify the program to remove barriers to success.
Determine if the activities assumed necessary for achieving outcomes were performed according to expectations.	If the necessary activities were performed and the initial conclusion was success, the activities probably are necessary for achieving outcomes and the program should be continued. If the necessary activities were performed but the initial conclusion was failure, the activities might not help achieve desired outcomes and the program process should be changed. If the necessary activities were not performed, program performance is inadequate and that corrective actions are needed.
Determine if there were unplanned benefits or unintended negative consequences.	If there are unplanned benefits, the evaluator should determine whether the benefits should be counted among planned outcomes in the future. If there are unintended negative consequences and program performance was as planned, this finding suggests that the process should be changed. If there are unintended negative consequences and program performance was not as planned, this finding suggests that staff or management methods may need to be changed.
Identify competing hypotheses. Judge whether the competing hypotheses could be as or more responsible for outcomes than the program activities or intervention. Serious competitors could require more investigation to reach a conclusion about the program's effectiveness.	List the results in the order of most damaging to most positive. Address the results at the top of the list first. Think of the most plausible explanations for each result. Debate the strengths and weaknesses of the competing explanations, and decide which explanations are most plausible.
Review findings and reach conclusions about merit.	Decide which conclusions should be used for making recommendations.
Draft the report and share with the staff of evaluand. Obtain their feedback.	If staff request modifications, decide which modifications to make and which aspects of the report are not modifiable.

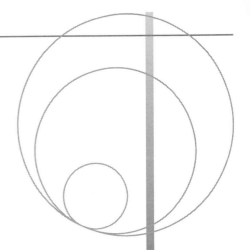

About the Authors

Phyllis M. Johnson, PhD, is a member of the faculty in the College of Health Sciences at Governors State University in south suburban Chicago, where she has developed and teaches workshops and courses in organizational development and quality systems, on which this manual is based. Prior to coming to Governors State, Johnson headed quality efforts at the Illinois Department of Children and Family Services. Before that, she was Director of Research and Management Information Systems and created the quality assurance unit at a Chicago-area child welfare agency. Johnson also worked directly with vulnerable populations as an instructor in general education development preparatory classes and a family service outreach worker, while also writing children's stories and word games.

Teresa L. Kilbane, PhD, is an Associate Professor at Loyola University Chicago, School of Social Work. She teaches research and policy courses to graduate and doctoral social work students. Her interests in research and program evaluation have led her to work and consultation in the area of quality assurance. For the past seven years, Kilbane has been a consultant for the Illinois Department of Children and Family Services. She co-authored this manual while working on the Organizational Development Demonstration Project. With Phyllis M. Johnson, she conducted quality assurance workshops with private child welfare agencies using curricula she developed based on material from the manual. She has also published several articles in the area of child welfare.

Laura E. Pasquale, MS, is a Consultant in Tallahassee, FL. Since 1990, Pasquale has worked in human services and the public sector, focusing on quality assurance, quality improvement, and performance evaluation and measurement. With experience in both the private, non-profit sector and state government, she applies her unique perspective on organizational dynamics in her work as an internal and external consultant. Pasquale is the former Bureau Chief for Quality Assurance and Accreditation at the Florida Department of Children and Families. She has served as a peer reviewer for the Council on Accreditation of Family and Children's Services, as Cochair of the Committee on Quality Improvement for the National Council on Research in Child Welfare, and as Chair of the Communications Committee for the Association of Rehabilitation Professionals in Computer Training. In addition to co-authoring this manual, Pasquale has assisted in peer review of field research for the Child Welfare League of America and is the author of several published articles.